FM 3-05.213 (FM 31-27)

Special Forces Use of Pack Animals

JUNE 2004

DISTRIBUTION RESTRICTION:
Distribution authorized to U.S. Government agencies and their contractors only to protect technical or operational information from automatic dissemination under the International Exchange Program or by other means. This determination was made on 2 April 2004. Other requests for this document must be referred to Commander, United States Army John F. Kennedy Special Warfare Center and School, ATTN: AOJK-DT-SFD, Fort Bragg, North Carolina 28310-5000.

DESTRUCTION NOTICE:
Destroy by any method that must prevent disclosure of contents or reconstruction of the document.

Headquarters, Department of the Army

Field Manual
No. 3-05.213

*FM 3-05.213 (FM 31-27)
Headquarters
Department of the Army
Washington, DC, 16 June 2004

Special Forces Use of Pack Animals

Contents

		Page
	PREFACE	iv
Chapter 1	MILITARY PACK ANIMAL OPERATIONS	1-1
	Characteristics	1-1
	Planning Considerations	1-2
Chapter 2	ANIMAL MANAGEMENT	2-1
	Mule Characteristics	2-1
	Donkey Characteristics	2-2
	Selection	2-5
	Animal Conformation	2-5
	Health and Welfare	2-9
	Feed and Water	2-13
	Feeding in Garrison	2-23
	Feeding in the Field	2-24
	Care of Forage	2-27

DISTRIBUTION RESTRICTION: Distribution authorized to U.S. Government agencies and their contractors only to protect technical or operational information from automatic dissemination under the International Exchange Program or by other means. This determination was made on 2 April 2004. Other requests for this document must be referred to Commander, United States Army John F. Kennedy Special Warfare Center and School, ATTN: AOJK-DT-SFD, Fort Bragg, North Carolina 28310-5000.

DESTRUCTION NOTICE: Destroy by any method that must prevent disclosure of contents or reconstruction of the document.

---

*This publication supersedes FM 31-27, 15 February 2000.

i

		Page
Chapter 3	**ANIMAL CARE AND TRAINING**	3-1
	Grooming	3-1
	Farrier Science	3-4
	Field Training	3-13
Chapter 4	**ANIMAL HEALTH MANAGEMENT**	4-1
	Animal Behavior	4-1
	Physical Examination	4-3
	First-Aid Supplies	4-7
	First-Aid Treatment	4-9
	Parasitic Infestation	4-14
	Diseases	4-15
	Hypothermia	4-19
	Heat and Sun Stress	4-21
	Immunization Schedule	4-22
	Medical Supply List	4-23
	Pharmacological Listing	4-24
	Euthanasia	4-24
Chapter 5	**PACKING EQUIPMENT**	5-1
	Selection of Equipment	5-1
	Packsaddles	5-1
	Halter and Packing Equipment	5-5
	Care of Equipment	5-10
	Fitting and Adjusting the Saddle	5-12
	Unsaddling the Animal	5-16
	Saddling With a Fitted Saddle	5-17
Chapter 6	**HORSEMANSHIP**	6-1
	Equipment	6-1
	Western and McClellan Saddles	6-4
	Riding Techniques	6-9
	Leading a Pack String	6-11
	Combat Considerations	6-14

		Page
Chapter 7	TECHNIQUES AND PROCEDURES	7-1
	Tying and Using Knots	7-1
	Wrapping Cargo With a Mantee	7-1
	Building Loads	7-11
	Special Weapons Loads	7-12
	Saddling	7-16
	Slings and Hitches	7-18
	The Pack String	7-23
	Campsites	7-25
	Transporting Sick and Wounded Personnel	7-29
Chapter 8	ORGANIZATION AND MOVEMENT	8-1
	Organization	8-1
	Duties and Responsibilities	8-1
	Movement Procedures	8-2
	Stream Crossing at Fords	8-4
	Crossing Unfordable Water	8-9
	Bivouac	8-15
Chapter 9	TACTICAL CONSIDERATIONS	9-1
	Security	9-1
	Cover and Concealment	9-2
	Actions on Contact	9-2
	Urban Environments	9-3
Chapter 10	LLAMAS AND OTHER ANIMALS	10-1
	Llamas	10-1
	Camels	10-5
	Dogs	10-6
	Elephants	10-8
Appendix A	WEIGHTS, MEASURES, AND CONVERSION TABLES	A-1
Appendix B	21-DAY PACK ANIMAL TRAINING PROGRAM	B-1
Appendix C	ANIMAL PACKING FORMATIONS	C-1
	GLOSSARY	Glossary-1
	BIBLIOGRAPHY	Bibliography-1
	INDEX	Index-1

Preface

Field manual (FM) 3-05.213 is a guide for Special Forces (SF) personnel to use when conducting training or combat situations using pack animals. It is not a substitute for training with pack animals in the field. This manual provides the techniques of animal pack transport and for organizing and operating pack animal units. It captures some of the expertise and techniques that have been lost in the United States (U.S.) Army over the last 50 years. Care, feeding, and veterinary medicine constitute a considerable portion of the manual; however, this material is not intended as a substitute for veterinary expertise nor will it make a veterinarian out of the reader. SF personnel must have a basic knowledge of anatomy and physiology, common injuries, diseases (particularly of the feet), feeding, watering, and packing loads to properly care for the animals and to avoid abusing them from overloading or overworking.

Though many types of beasts of burden may be used for pack transportation, this manual focuses on horses, mules, donkeys, and a few other animals. One cannot learn how to pack an animal by reading; there is no substitute for having a horse or mule while practicing how to load a packsaddle for military operations. However, the manual is useful for anyone going into an environment where these skills are applicable.

The most common measurements used in pack animal operations are expressed throughout the text and in many cases are U.S. standard terms rather than metric. Appendix A consists of conversion tables that may be used when mission requirements or environments change.

The proponent of this publication is the United States Army John F. Kennedy Special Warfare Center and School (USAJFKSWCS). Submit comments and recommended changes to Commander, USAJFKSWCS, ATTN: AOJK-DT-SFD, Fort Bragg, NC 28310-5000.

Unless this publication states otherwise, masculine nouns and pronouns do not refer exclusively to men.

Chapter 1

Military Pack Animal Operations

Since the deactivation of the pack transport units after the Korean Conflict, the Army has relied on air and ground mobility for transporting personnel and equipment. Today and throughout the operational continuum, SF may find themselves involved in operations in rural or remote environments, such as Operations UPHOLD DEMOCRACY or ENDURING FREEDOM, using pack animals.

SF personnel must conduct a detailed mission analysis to determine the need for pack animals in support of their mission. Military pack animal operations are one of the options available to a commander to move personnel and equipment into or within a designated area of operations (AO). Pack animal operations are ideally suited for, but not limited to, conducting various missions in high mountain terrain, deserts, and dense jungle terrain.

Personnel must have a thorough understanding of all the factors that can impact military pack animal operations. The objective of this chapter is to familiarize the reader with military pack animal operations and to outline the planning considerations needed to successfully execute them.

CHARACTERISTICS

1-1. Commanders use military pack animal operations when the AO restricts normal methods of transport or resupply. Animal transport systems can greatly increase mission success when hostile elements and conditions require the movement of combat troops and equipment by foot.

1-2. The weight bearing capacity of pack animals allows ground elements to travel longer distances with less personnel fatigue. The pack train can move effectively and efficiently in the most difficult of environments with conditioned animals, proper/modern equipment, and personnel with a moderate amount of training in handling packs. The pack detachment, without trail preparation, can traverse steep grades and heavily wooded areas, and can maintain acceptable speeds over terrain that is not mountainous, carrying 35 percent of their body maximum (150 to 300 pounds [lb]). This amount should be decreased for loads that are prone to excessive rocking as the animal walks (for example, top-heavy loads and bulky loads). This capability continues indefinitely as long as the animals receive proper care and feed. In mountainous terrain, with no reduction in payload, the mule or horse can travel from 20 to 30 miles per day.

1-3. The success of pack operations, under extreme weather and terrain conditions, depends on the selection and training of personnel and animals.

Personnel involved in pack animal operations require extensive knowledge of pack animal organization and movement, animal management, animal health care, pack equipment, and load planning. Planning the use of pack animals is not a simple task, nor is it always a satisfactory solution to a transportation problem.

1-4. When used correctly, pack animal operations give commanders another means to move Special Forces operational detachments (SFODs) and influence the battlefield. The skills and techniques used in pack animal operations are applicable to all SF missions.

PLANNING CONSIDERATIONS

1-5. Successful pack animal operations depend on thorough mission planning, preparation, coordination, and rehearsals. Initial mission planning should include a mission, enemy, terrain and weather, troops and support available—time available and civil considerations (METT-TC) analysis to assist in determining whether or not to use pack animals in a mission. Analyzing the METT-TC questions (Table 1-1, pages 1-2 and 1-3) will enable the executing element to consider all factors involved when using pack animals in their mission profile.

Table 1-1. METT-TC Analysis

Factors	Questions
Mission	<ul><li>What is the duration of the mission?</li><li>What type of terrain does the operational area comprise and at what altitude is it located?</li><li>Does the mission profile require a lot of movement? Is the projected rate of foot movement feasible?</li><li>Is the operational area conducive to pack animal use? Is the mission time-critical?</li></ul>
Enemy	<ul><li>How will the enemy threat, capabilities, disposition, and security measures affect the mission?</li><li>Does the enemy use pack animals?</li><li>Does the enemy have a similar capability to detect or interdict conventional infiltration methods?</li></ul>
Terrain and Weather	<ul><li>Is the terrain conducive to pack animal operations?</li><li>Does altitude prohibit or restrict pack animal operations?</li><li>Does seasonal bad weather prohibit or restrict pack animal use?</li><li>Does the detachment have experience navigating pack animals in limited visibility conditions?</li></ul>

Table 1-1. METT-TC Analysis (Continued)

Factors	Questions
Troops and Support Available	<ul><li>Does the detachment have the training and experience to successfully execute pack animal operations?</li><li>Are pack animals available for training and rehearsals?</li><li>What types of pack animals are available in the operational area?</li><li>What special equipment is required to conduct pack animal operations?</li><li>What is the anticipated duration of the operation?</li><li>Are there areas for the animal to graze or forage?</li><li>Does the detachment have the means to infiltrate the required equipment into the AO?</li><li>Does the equipment require special rigging? Does it have special handling and storage requirements?</li><li>Is the detachment going to use the local pack equipment? Does the detachment know about local pack equipment?</li><li>Does the detachment need to hire a local handler to pack the animals? Will the handler travel far from home?</li></ul>
Time Available	<ul><li>Is time available for the detachment to plan and rehearse pack animal use before mission execution?</li><li>Will time be available on the ground for the detachment to rehearse packing the animals?</li><li>Will there be time to acquire local equipment and feed, and to inspect animals if needed?</li></ul>
Civil Considerations	<ul><li>Can the operation be executed clandestinely so that the civilian populace is unaware of it?</li><li>If the operation is compromised, what will be the repercussions to the local populace?</li><li>If the detachment is receiving support from the locals, is there a risk of reprisals against them?</li></ul>

1-6. A thorough METT-TC analysis concentrating on the above questions pertaining to pack animal operations will determine if pack animal use is appropriate. The detachment must then complete the remainder of the mission planning process.

Chapter 2

Animal Management

The survivability of a pack animal detachment and its ability to successfully complete a mission depend on the animals and their management. Historically, animals of all types and sizes have been successfully used for pack transportation throughout the world. Animals indigenous to the AO are usually more effective than imported animals. Although native animals may be smaller and not ideally proportioned, they are acclimated to the environment, generally immune to local afflictions, and accustomed to the native forage. Nevertheless, any animal locally procured needs to be thoroughly inspected for disease and physical soundness. Animal management entails selection, feed, and feeding along with stable management. Employing an untrained pack master, making poor animal selections, and improperly feeding the pack animals could prove disastrous for the detachment.

MULE CHARACTERISTICS

2-1. Mules are the hybrid product of a male donkey and a female horse (Figure 2-1, page 2-2). Male mules are called johns and female mules are called mollies or mare mules. Mollies are a cross between male donkeys and Belgium horse mares. Mollies generally have a gentler disposition than johns. Intelligence, agility, and stamina are all characteristics of mules. These qualities make mules excellent pack animals. Unlike horses, which carry about 65 percent of their weight on their front legs, mules carry 55 percent on their front legs. This trait makes them very well balanced and surefooted in rugged terrain.

2-2. Some people think stubbornness is a mule characteristic—stubborn as a mule! Mules are intelligent and possess a strong sense of self-preservation. A packer cannot make a mule do something if the mule thinks it will get hurt, no matter how much persuasion is used. Therefore, many people confuse this trait with stubbornness. Mules form close bonds with horses, especially mares. The bond is so close that mules willingly follow a mare. That is why a mare will usually be wearing a bell leading a string of mules. A wrangler, or mule skinner, can usually control an entire pack string simply by controlling the bell mare. At night in the backcountry, mule skinners can picket the bell mare and turn the mules loose. The mules will disperse and graze freely, yet remain close to the mare. Environmental impacts are reduced and the mules are easy to gather in the morning.

2-3. Young mules are naturally and easily startled but if treated with great patience and kindness can easily be broken in. All harsh treatment of any kind must be avoided or could prove to be fatal to successful training.

FM 3-05.213

Figure 2-1. Mule

DONKEY CHARACTERISTICS

2-4. The first thing a person thinks of when a donkey comes to mind is what? Big ears? Or maybe a short whisk broom tail? Figure 2-2, page 2-3, shows the donkey's features, which help him succeed and survive in a harsh environment. Donkeys vary in size and provide different levels of transport. A detachment can use any of the following-sized donkeys:

- Miniature: up to 36 inches.
- Small standard: 36 to 40 inches.
- Standard: 41 to 48 inches.
- Large standard: 48 to 56 inches.
- Mammoth jack stock: 54 inches and up for jennets (females); 56 inches and up for jacks (males).

2-5. Donkeys evolved in the desert. Because food was usually scarce, high concentrations of donkeys in one area were not possible. Donkeys still have to drink water daily, but due to desert adaptations, their bodies do not waste or lose moisture as readily as does a horse. The donkey's body extracts most of the moisture from his own feces and does not need to sweat as much as a horse would (a donkey does not have large muscle mass to always have to keep cool), which makes him a better water conservationist.

2-6. The donkey's mighty bray allowed even widely spaced donkeys to keep in contact or define their territories. Their big funnel-shaped ears could catch the distant calls and maybe help dissipate some hot desert heat. Their ears also serve as a visual communication system, telegraphing danger or asinine moods. They punctuate their "ear-ial" code messages with tail swishes, body language, and of course, grunts and moans.

Figure 2-2. Donkey

2-7. Other special characteristics of donkeys are tough, compact hooves that can handle sand and rock, woolly hair to insulate from desert heat and cold, and a lean body mass that is fuel-efficient and easily cooled, yet very strong and enduring. They also have a digestive system that can break down almost inedible roughage while at the same time extracting and saving moisture in an arid environment. Donkeys have only five lumbar vertebrae compared to most horses, which have six. They also generally have upright, sparse, spiky manes with no forelock.

2-8. Donkeys come in various colors, but the most common (for standards and miniatures) is the mouse gray called gray dun. There are spotted donkeys, white donkeys, various shades of brown that breeders refer to as "chocolate," black donkeys, sorrel donkeys, and even pink donkeys, which have light, red hair mixed with a gray dun coat giving the illusion of pink. There are also various roan and frost patterns. Donkeys come with or without a cross, leg stripes, or collar buttons. Most have white muzzles, eye rings, and light bellies. Mammoth jack stock tend to not have crosses and usually are seen as black, red, red roan, blue, blue roan, and spotted combinations, to name a few.

2-9. Regardless of the packaging and through the ages, different countries bred the donkey in whatever form they needed (the donkey had many uses then, as it still does today—riding, packing, draft work, creating mules), so donkeys now come in any size (from 64-inch-plus mammoths to 28-inch-or-less miniatures), shape, coat color, and length of hair. But inside all is the same gentle, calm, and slightly mischievous soul.

2-10. Donkeys are cautious of changes in their environment. Donkeys have a strong sense of survival. If they deem something as dangerous, they will not do it. It is not stubbornness—it is Mother Nature, and they are smart enough to know when they cannot handle something. A handler should never lose his temper or use brute force to accomplish a task because the donkey will then fear his handler for life. Yet, with trust and confidence in their handlers, donkeys will go along with what tasks are necessary and accomplish the mission.

2-11. Donkeys, because of the rugged terrain that they evolved in, could not just run away from danger in absolute panic. Running without caution simply placed them in further peril. Natural selection weeded out the less intelligent, so that now donkeys generally will freeze when frightened or run a little way then stop to look at what startled them. (This instinct to freeze rather than flee is what is so desired in the mule, along with the donkey's stamina and intelligence.) Donkeys will also naturally attack canines to protect themselves and their young (or if properly conditioned, a donkey will also protect sheep and goats).

2-12. This strong, calm, intelligent worker that does not tend to run away in terror after being spooked and has a natural inclination to like people adds up to an animal that is easy to take care of and easy to work.

2-13. Table 2-1, pages 2-4 and 2-5, explains the differences in the makeup of the equine family.

Table 2-1. Equine Comparisons

Features	Donkey	Mule	Horse
Ears	Long.	Medium (depends).	Short.
Muzzle	Normally white.	Normally tan.	Usually same color as entire face; bare in the summer.
Eye Rings	White.	Lacking, though shape of orbital eye sockets usually distinctive in the mule.	Lacking.
Chest	Smooth muscling over chest, not divided.	Smooth muscling over chest, normally not divided.	Divided breast muscles over thickly muscled chest.
Belly	Usually white.	Color same as body.	Color same as body.
Vertebrae	Usually has 5 lumbar vertebrae.	Not documented; usually depends on genetics from parents.	Normally has 6 lumbar vertebrae (some barbs and mustangs have only 5).
Head	Wide between eyes and crown with a much bigger jaw.	Usually intermediate, but normally larger than a horse's; has a wider brow and jaw.	Usually petite in comparison to long-eared counterparts.
Withers	Usually lacking distinctive horse withers; can make the back appear even longer.	Usually lacking withers in comparison to the horse.	Average horse has a 2- to 4-inch prominence above and forward of the back line.

Table 2-1. Equine Comparisons (Continued)

Features	Donkey	Mule	Horse
Hip and Hindquarters	Normally steeper than a horse; croup can be somewhat peaked.	Varies depending on bloodlines; normally reflects the smooth muscling of the donkey but with a larger muscle mass. Top lines can reflect either parent.	Normally more muscled than mules and donkeys; muscle groups are defined depending on breed.
Chestnuts	Have only front, which are flat, soft, and leathery.	Can follow either parent or be a mixture of the two.	Normally one on each leg; thick, horny protrusions that grow and sometimes require trimming.
Tail	Has longer coarse tail hairs on the bottom third, short hairs at the top third.	Normally tailed like a horse, though often not as full in body, with stiffer hairs.	Normally full and long, depending upon breed.
Hooves	Normally steeper and more boxy than mules and horses; frog is more developed with a thicker sole. Hoof angles average about 65 degrees.	Usually steeper than a horse and boxier rather than oval.	Normally quite oval with walls that spread out; frog is less developed; hoof size is larger in comparison to body size than the donkey. Hoof angles average around 55 degrees.
Gestation	Average 365 days.	Not applicable.	Average 342 days.

SELECTION

2-14. Mobility and effectiveness of the pack animal detachment depend largely on the selection and training of the pack animals. The pack animal, regardless of its color, breed, or size, should have a friendly disposition, a gentle nature, and no fear of man. It should be willing to travel under a load and be sure-footed. Large, draft-type horses usually are not agile and do not make good pack animals. The ideal pack animal should be 56 to 64 inches (14 to 16 hands) tall. (Since one hand equals about 10 centimeters [cm], the metric equivalent is 140 cm to 160 cm.) It must be tough, compact, sturdy, and well-formed. Figure 2-3, page 2-6, shows specific parts of an ideal pack animal (horse). For additional information and a more detailed illustration of the pack animal, see *Care and Feeding of the Horse*, Lon D. Lewis, ISBN 0-68-304967-4.

ANIMAL CONFORMATION

2-15. Conformation is a term used to describe what a good pack animal should look like with regard to size, shape, height, weight, and dimensions. The single most important conformation factor is good body condition. Thin animals are poor candidates for many reasons. They should be avoided if they do not gain weight after being given adequate nutrition. Thin animals do not tolerate the rigors of long pack trips since they have little time to eat or graze. They easily develop saddle or pack sores on their backs because of the prominence of bones. The lack of body condition may be due only to poor nutrition. However, it may be due to more serious problems such as bad teeth or chronic infection. A pack animal should be calm and not easily spooked, easy to work around and not stubborn, and should be able to socialize well with other pack animals so that it is not too timid or too aggressive.

FM 3-05.213

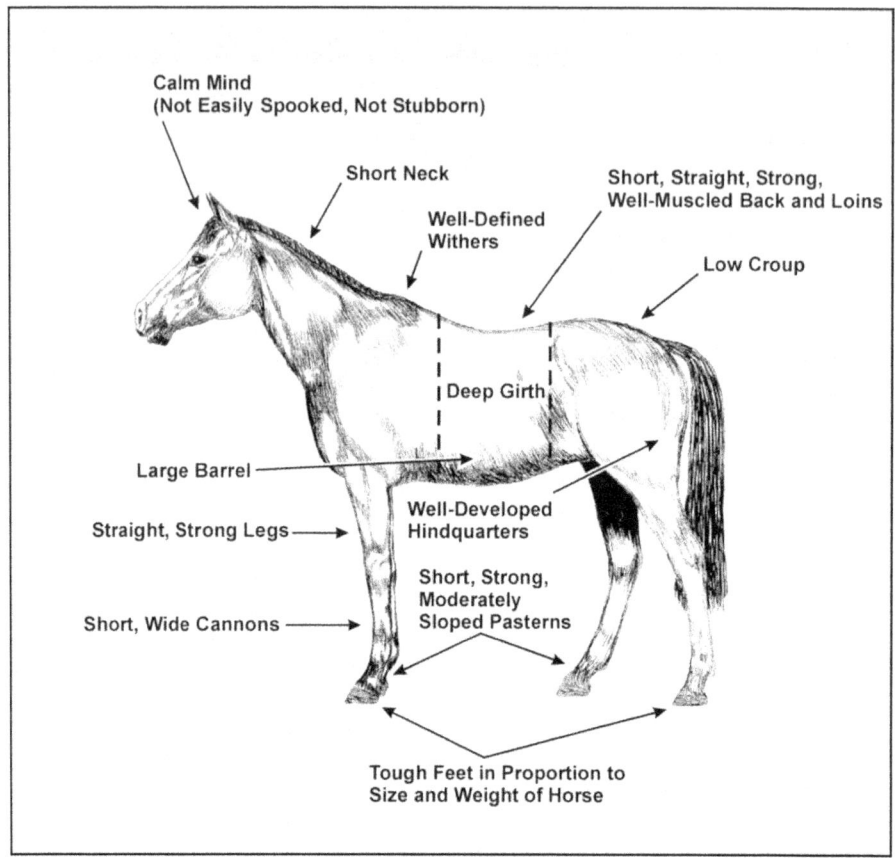

Figure 2-3. Animal Conformation

GOOD CONFORMATION

2-16. Good conformation is marked by—
- A steady, intelligent mind.
- A short, strong neck.
- Well-defined withers that are not too high or too low.
- A large muscular chest.
- Short, straight, strong, well-muscled back and loins.
- A low croup.
- A deep girth and large barrel to accommodate big lungs.
- Well-developed hindquarters.

- Straight, strong legs.
- Short, wide cannon bones.
- Short, strong pasterns that have a moderate slope.
- Tough hooves in proportion to size and weight of animal.

BAD CONFORMATION

2-17. Bad conformation is marked by—
- A poor disposition.
- Withers that are too high or too low.
- A hog back or swayback.
- A shallow girth or small barrel.
- Underdeveloped muscle groups.
- Long, spindly legs.
- A general history of poor health.
- Advanced aging.

2-18. The **head** should be proportioned to the neck that supports it. A big head on a long, weak neck does not give the animal a desirable pendulum for balance. A well-formed head is usually a sign of good breeding. The animal's eyes should be clear and free from any cloudiness in the cornea or fluid within the eye. There should be no drainage of tears over the lower eyelids. The white sclera around the eyes should be white with no hint of infection or yellow discoloration (jaundice). The pink conjunctiva (clear membrane that goes over the white of the eye) under the eyelids should be pink, not red, and should have no hint of infection. The animal must also have good vision to pick its way through rough or rocky terrain.

2-19. The **teeth** should match to form a good occlusion. They should not have sharp points or hooks. Although a person skilled in floating teeth can remove these sharp points, they may soon return due to improper occlusion. **Good teeth, lips,** and **tongue** are important to the animal's survivability by allowing it to eat and drink properly. The animal needs good lips to find and pick up grain and water since it cannot see the end of its nose. The junction of the head and neck should be strong without a meaty appearance.

2-20. The **neck** should be proportioned to the animal's body and not too long. The animal needs strength in the neck to use its head as a pendulum and keep its balance in adverse terrain.

2-21. On a pack animal, the **withers** should be prominent, but not so high the packsaddle will rub or ride on it. Flat, rounded, or mutton (flat and thick) withers should be avoided because the packsaddle will not ride well on the animal. Attention is given to the height of the withers in relation to the height of the croup. An animal with withers lower than the croup has greater difficulty carrying a load because the load will constantly ride forward causing greater strain on the shoulders and forelegs. In contrast, an animal with extremely high withers and a low croup has difficulty with the load sliding back into the hip, leaving the possibility of the breast collar impeding movement and breathing.

2-22. The **shoulders** of a pack animal should be long, deep, and sloping to provide a larger surface for bearing the weight of a pack load when the animal is ascending hilly terrain. The **chest** should be muscular and deep but not too broad.

2-23. The **legs** should be straight, well-muscled, and free of any bulging joint capsules. The detachment should avoid animals that have any of the following characteristics on any legs or hooves:

- Club feet (foot and pastern axis of 60 degrees or more).
- Long, sloping pasterns (foot and pastern axis less than 45 degrees in front or less than 50 degrees in back).
- Short, upright pastern (foot and pastern axis more than 55 degrees in front or more than 60 degrees in back).
- Thin hoof walls.
- Thin sole.
- Buttress foot (swelling on the dorsal surface of the hoof wall at the coronary band).
- Bull-nosed foot (dubbed toe); unilateral contracted foot.
- Toed-in (pigeon-toed) or toed-out. (Normal animals will toe-in or toe-out if not trimmed correctly. The detachment should avoid animals that continue to toe-in or toe-out after trimming to ensure the bottom of the hoof [inside heel to outside heel] is perpendicular to the median plane of the body.)
- Any hoof that is sensitive to moderate pressure applied by a hoof tester to the toes, sole, frog, or across the heels.

2-24. There are also specific defects that pertain only to the front and back legs of the animal. The detachment should avoid animals with the following defects:

- Front legs:
  - Open-knees (enlargement of the distal growth plate of the radius and corresponding enlargement in the knee).
  - Calf-knees (backward deviation of knees).
  - Buck-knees (forward deviation of knees).
  - Offset knees or bench knees (knee sits too far medially over lower leg rather than sitting centered over lower leg).
- Rear legs:
  - Cow hooks (knock-knees).
  - Sickle leg (excessive angle of the hock).

2-25. The **feet** are a critical factor in the animal's ability to perform, stay physically sound, and endure the hardships of packing. Ideally, the horse should stand with its feet at a 45- to 50-degree angle to the ground. The size of the foot should be in proportion to the size of the animal. Small feet are often brittle and do not have the base to support a heavy load or absorb concussion. Large feet could cause the animal to be clumsy and awkward. Mules have a tendency to have smaller feet, but this fact does not present a problem. The sole should be slightly concave and the frog prominent, flexible, and tough. Again, when viewed from the front, an animal that is toed-out should be avoided.

2-26. The **girth** should be deep from the withers to the floor of the body, and the body should be wide and flat. This size indicates ample space for vital organs, such as the heart and lungs. The barrel should be large. A large barrel shows a good spread of the ribs that, in turn, give a good load-bearing surface on top.

2-27. The **back** should be short, strong, and well-muscled. A short back is better equipped to carry a load without sagging. Horses with one less vertebra than others would be good selections. The backbone should be prominent. A pack animal with a rounded back and ill-defined backbone is difficult to pack so that the load rides properly. Chances are good the load will slip or roll, and the detachment will waste valuable time repacking the load during movement.

2-28. The **loin** should be of moderate length, well-muscled, and broad. A long loin will cause weakness at that point. The **croup** should be low and of moderate width and slope. The hindquarters should be strong and well-developed to provide power to the animal.

CONFORMATION DEFECTS

2-29. In the selection of a pack animal, the above criteria are ideal, but many serviceable pack animals have defects in their conformation and still perform well. Nevertheless, it is better in the long run to avoid animals with many defects. Personnel should try to ensure that the larger animals carry the heaviest loads, and the gentle, experienced animals carry the fragile, easily breakable items. The smaller animals or animals with certain conformation defects should be tasked with carrying the light and not-so-fragile loads, such as food for animals and Soldiers. Bigger (size and weight) is not necessarily better. A load of 100 to 150 pounds is big enough for most packers. Heavier loads risk injury to an animal unless it is exceptionally well proportioned. The detachment probably cannot use the extra capacity anyway. The pack handler will have to lift the load in the field, and a 100-pound load is usually about as much as a person can properly lift and position by hand on a packsaddle. Between fourteen and fifteen hands (4 feet 7 inches to 5 feet 2 inches) is a good range for pack stock. (**Note:** One hand equals 10.2 cm or 4 inches.) Even men more than 6 feet tall can have difficulty loading animals that are higher.

HEALTH AND WELFARE

2-30. The health and welfare of the pack animal is a major concern of animal handlers in garrison and in the field. Each individual provides for the welfare of pack animals. Whole pack trains can be lost and missions compromised because of poor animal care. Implementing animal care will be distinctly different for field and garrison conditions even though the requirements and desired end points are the same.

FM 3-05.213

FIELD MANAGEMENT

2-31. Rarely will a packer have the luxury of shelter or even corrals to hold animals in when packing. If fencing or corrals are not available, some means of limiting movement is necessary. The packer can limit movement by tying long "stake-out" lines to trees, using the high line or by using auger devices secured into the ground. These methods permit animals to graze in defined areas and provide windbreaks during cold winds. However, there are some disadvantages. The animals must be moved every few hours as they graze the available forage. Sometimes they become entangled in the stake-out lines or entangle the line in brush or rocks. Some animals panic at loud noises (for example, thunder) and snap the line or pull the auger out.

2-32. The packers can hobble the animals by tying their front legs together with just enough space between their legs to take small steps but not run. They can secure the hobbles around the pastern or above the knees. Pastern hobbles (American hobbles) are easier to apply and maintain but can cause some animals to become entangled in brush or rock. Hobbles above the knees (Australian hobbles) are more difficult to apply and maintain but prevent entanglement with brush and rocks. Packers must slowly acclimate animals to hobbles in garrison before using the hobbles for an extended time on pack trips. Figure 2-4 shows two types of hobbles that the handler can use for various purposes.

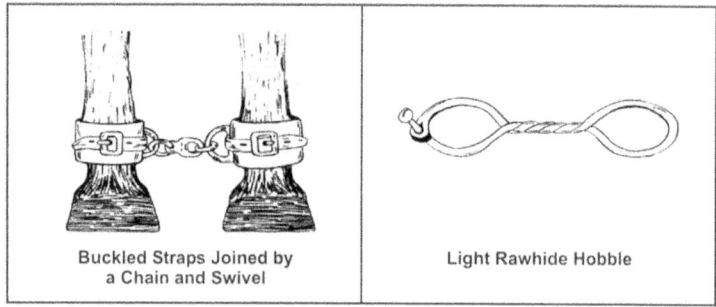

Buckled Straps Joined by a Chain and Swivel

Light Rawhide Hobble

Figure 2-4. Two Types of Hobbles

2-33. The animals must have as much access to good forage as possible. This need means that packers should include probable grazing areas in their movement schedules. Even with optimal grazing, pack animal detachments take additional supplemental feed (in the form of grain) because the animals do not have enough time to graze and rest after carrying packs for much of the day. Also, the grazing is often less than optimal and the animals have to range over considerable distances to obtain adequate nutrition.

NOTE: Pack animal detachments can also consider using portable electric fencing. However, a few animals are not deterred under normal conditions, and most animals are not deterred if spooked by thunder and lightning.

SHELTERED HOUSING FACILITY (STABLE BUILDING)

2-34. In an unconventional warfare (UW) scenario, the pack animal detachment will probably not have the luxury of a permanent or semipermanent stable facility. However, most of the field routines may be applied in stables by support personnel before the mission or after the unit has completed their operation.

2-35. The detachment commander has command responsibility for stable management and the training of his Soldiers. However, subordinate leaders are directly responsible for stable management and the stable routine.

2-36. Stable management includes the supervision and maintenance of the stables and other facilities. The subordinate leaders will ensure the grounds and buildings in the stable area are kept as clean and sanitary as available time and labor will allow and that the grounds are reasonably level and well drained. They will also ensure the animals are well-groomed, properly shod, and free of injuries and diseases. Since a large number of animals may be involved in a pack animal detachment, subordinate leaders should keep records at the stables on all the assigned animals.

2-37. The design and construction of a stable facility may be limited to the materials at hand. Regardless of the materials used, the stable should provide adequate shelter, good ventilation, and few maintenance requirements. The stable building should provide ready access to the corrals and storage for feed and packing equipment.

STALLS

2-38. Stalls vary in size depending on the average size of the animals, amount of time the animals are expected to spend in the stall, and available space. A 12- by 12-foot stall allows freedom for a large horse or mule to maintain fair physical condition during long periods of idleness while confined to stables. Stalls of this size are normally used for recuperation, foaling of mares, and protection in extreme climates. A 10- by 10-foot stall is normally satisfactory when animals are in stables only for feeding and rest. To reduce waste in feeding hay and grain, the stalls should be equipped with hayracks and feed boxes. Stall walls must be free of sharp or rough projections and unfinished woodwork. A major concern when dealing with animal care is the construction and maintenance of stall floors. The floor should be level and have good drainage. It should also be resilient to help maintain a healthy condition of the animals' feet and legs. The floor should also be nonabsorptive for cleanliness and sanitation. Earthen floors composed of clay are satisfactory but require continual work to clean, level, and smooth. Rough-finished concrete provides the best type of floor because it is sanitary, is easy to clean, and requires little maintenance. However, a concrete floor having little resilience must be covered with a bedding of straw or hay for cushion. Wooden floors, even if impregnated, are not desirable since they are slippery and unsanitary. Because of their porosity, they cannot be adequately disinfected such as would be necessary during a Salmonella outbreak. Regardless of the floor chosen, a good bed contributes to the comfort and efficiency of the animals. A clean, comfortable bed will induce the animal to lie down and get better rest. It provides a soft surface that will prevent bruising or abrasion of elbows, hocks, and other body parts. It also provides insulation for the body and a comfortable surface for the animal to stand.

2-39. Horses and mules are herd animals and do not thrive in stalls. This confinement can lead to behavioral problems that would not be experienced when the animals are kept in a field environment.

STORAGE

2-40. Storage facilities provide protection and security for feed and tack. They should also be convenient to the stables and corral. Feed storage should provide protection from rodents, water, and any loose animals that may overeat. When possible, personnel should stack feed on pallets (best), plastic sheets (second best), or boards so that it is not in direct contact with the ground. Feed should be stacked at least 6 inches from walls to discourage rodents. Rodents prefer narrow spaces to avoid predators, and the extra distance decreases the frequency of rodent visits. When covering feed outside, there should be no pinpoint holes in the cover because they are just enough to cause spoilage if it rains. The detachment should also arrange to minimize time and personnel necessary for feeding. A tack room should be planned for each stable and partitioned from the stall area. It should have facilities for inspection, cleaning, preserving, repair, and storage of all pack and riding saddles, bridles, halters, panniers, and accessories for the detachment.

CORRALS

2-41. The pack animal detachment should provide corrals for the animals to move freely and exercise when they are not in use or in the stalls. The corrals should be close and easily accessible to the stables, be well drained, and provide good footing. Personnel should fill and level, as much as possible, any depressions and heavily traveled areas where water can collect. They should set fences at a sufficient height and strength to ensure the safekeeping of the animals. Fences and gates should be of a height above the base of the horse's neck to prevent escape or injury in attempts to escape. The fence should be free of sharp or rough projections, exposed nails, and edges prone to splintering with pressure or rubbing. The animals should have water troughs or tanks to allow them free access to water. The containers should be large enough to allow the watering of several animals without congestion. Individual feed boxes or buckets are much better than shared bunks to control contagious diseases and minimize fighting. If feed bunks are used, they should provide at least 3 feet of bunk space per animal with no more than four animals per bunk to decrease fighting over feed and to prevent one dominant animal from eating more feed than the others. Handlers should establish hitching posts or a picket line to groom or pack the animals. The footing at the picket line must be strong enough to sustain heavy use. The line should be established on high ground so water (even if dry season), urine, and feces do not accumulate around the picket line. Also useful is a foundation of stone cover with coarse (but smooth) gravel or sand. Personnel should try to avoid fine gravel or coarse gravel with rough edges as these can lodge in the frog or sulcus. If there is no natural shelter from the sun and bad weather, the handler should provide shelter for the animal's protection.

SOCIAL DOMINANCE

2-42. Horses (and mules) develop a social hierarchy or "pecking order." All animals in the herd respect this hierarchy. Fighting and injury prevail until the hierarchy is established. Once the hierarchy is established, fighting decreases. The hierarchy can change as new animals are introduced or as dominant animals age or become debilitated.

2-43. The handler should let the animals establish their hierarchy but manage it so as to minimize the injuries as the hierarchy forms. Introducing new animals to the herd with a wooden fence (or other similar barrier) separating them will help in managing the herd. This method decreases the chances of kick and bite injuries as subordinate animals try to escape.

2-44. The hierarchy can be very helpful if understood and used. The dominant animals fight to be at the front of the pack line and fight to eat first. Handlers should identify the dominant and subordinate animals and place them in the pack line according to the hierarchy. Likewise, the animals should be fed in the same order to reinforce the existing hierarchy. Attempting to alter the hierarchy will likely result in an even greater effort of the horses or mules to reestablish the hierarchy, leading to fighting and injury.

SANITATION

2-45. Sanitizing at the stables and in the field is a continuous process for maintaining the health of animals and personnel. Stables and corrals must be kept clean to reduce the breeding of flies, which is one of the most serious sanitation problems that lead to disease and infection. The most effective countermeasure to reduce flies is to reduce moisture and harborage. Good drainage and the removal of wet or soiled bedding help control moisture. Eliminating areas favorable to fly breeding and hatching—that is, moist, dark areas with organic material—controls fly harborage. Personnel should also brush out all feed boxes daily and scrub them monthly. Individual watering buckets or troughs are preferred. If this is not possible, the water trough should be drained and cleaned weekly. Only the animals belonging to the detachment should drink from these troughs. Most importantly, personnel should ensure birds cannot defecate in the troughs, and rodents cannot swim in them. Any animals suffering from a communicable disease should be watered from a bucket that is thoroughly cleaned and disinfected after each use and not shared with other animals.

FEED AND WATER

2-46. The health, condition, and effectiveness of a pack animal directly relate to the type and amount of food being consumed. The animal handler determines the amount and type of food by the amount and type of work to be performed. The working animal needs more concentrates (proteins, minerals) in its diet to supply fuel for energy, replacement of tissue, and maintenance of condition. An idle animal does not require as many nutrients in its diet. Personnel in the pack animal detachment need to have a basic knowledge of feed grains and roughage, their characteristics, and their geographical availability. This information is also critical for operational planning.

FEED REQUIREMENTS

2-47. The body requires food for growth, repair of body tissue, and energy for movement. The body also needs food to maintain temperature and energy for such vital functions as circulation, respiration, and digestion. Protein from feeds such as oats, barley, corn, and bran provide for body growth and repair of tissue. Minerals from feeds such as grass, hay, bran, and bone meal are needed for healthy bones. Carbohydrates and fats from feed such as corn, wheat, rye, and oats produce heat and energy or are stored as fat and sugar as an energy food reserve. Bulk feeds such as hay, grass, bran, and oats are necessary for digestion. Most foods, with the exception of hay or good forage, do not contain all the necessary nutrients for horses and mules; therefore, foods are combined to obtain the desired nutritional value. Oats are the best grain feed for stabled animals. For animals on pasture, natural grasses come closest to providing all required nutrients. The nutritional value of feed is measured in terms of the amount and proportions of digestible nutrients it supplies.

2-48. A ration is the feed allowed one animal for 24 hours, usually fed in two portions, morning and evening. The components of a ration depend upon the class and condition of the animals, the work being done, the variety of available foods, the kind of shelter provided, the climate, and the season. Feed must be selected and combined, proportionately, to form a balanced ration that consists of proteins, carbohydrates, fats, minerals, and vitamins. Pack animals cannot thrive on concentrated foods alone. They need both forage and concentrates, but the ratios need to be controlled as explained below. Quantities of feed in one ration vary depending on the amount of idle time the animals have, the work being performed, and the availability of the feed. Insufficient feed—particularly bulky feed—causes loss of conditioning, general weakness, and predisposes an animal to disease. Too much feed can be wasteful and cause an animal to be overweight. An animal that is too heavy will tire prematurely, suffer heat stress, and possibly develop laminitis (lameness caused by swelling of the feet).

GUIDELINES FOR FEEDING

2-49. All pack animal handlers should adhere to specific principles for optimal feeding. The following paragraphs explain each of these principles.

2-50. **Feed to Maintain a Body Condition Score (BCS) of 5.** Table 2-2, page 2-15, clarifies the differences seen with different BCSs. It is important for animals to maintain a BCS of 5 for several reasons. If animals are too thin, they will—

- Be prone to saddle sores because the bones (especially the withers) will protrude.
- Have less energy and be sluggish.
- Have less body reserve to sustain them if rations are unavailable.
- Be more susceptible to infections and contagious diseases.

2-51. Handlers must also be aware of the opposite effect. If animals become too heavy, they will—

- Be wasting feed that may be needed later.
- Be predisposed to heat stress.

- Be predisposed to premature exhaustion.
- Cause their packs not to fit well.
- Be prone to laminitis and other problems of lameness.

Table 2-2. Horse's Appearance Associated With Dietary Energy Intake

BCS	Descriptions
2—Very Thin	Animal emaciated; slight fat covers base of spinous process; transverse processes of lumbar vertebrae feel rounded; spinous processes, ribs, tailhead, tuber coxae, and tuber ischii prominent; withers, shoulders, and neck structure faintly discernible.
3—Thin	Fat buildup about halfway on spinous process; transverse process cannot be felt; slight fat cover over ribs; spinous processes and ribs easily discernible; tailhead prominent, but individual vertebra cannot be identified visually; tuber coxae appear rounded but easily discernible; tuber ischii not distinguishable; withers, shoulders, and neck not obviously thin.
4—Moderately Thin	Slight ridge along back; faint outline of ribs discernible; tailhead prominence depends on conformation, but fat can be felt around it; tuber coxae not discernible; withers, shoulders, and neck not obviously thin.
5—Moderate	Back is flat (no crease or ridge); ribs not visually distinguishable but easily felt. Fat around tailhead slightly spongy; withers appear rounded over spinous process; shoulders and neck blend smoothly into body.
6—Moderately Fleshy	May have slight crease down back; fat over ribs spongy, fat around tailhead soft; fat beginning to be deposited along the side of withers, behind the shoulder, and along the side of the neck.
7—Fleshy	May have crease down back; individual ribs can be felt, but there is a noticeable filling between ribs with fat; fat around tailhead soft; fat deposits along withers, behind shoulders, and along the neck.
8—Fat	Crease down back; difficult to feel ribs; fat around tailhead very soft; area along withers filled with fat; area behind shoulders filled with fat; noticeable thickening of neck; fat deposits along inner thighs.

Sources: *Care and Feeding of the Horse*. Lon D. Lewis. 2d Ed., Williams and Wilkins, 1996. *Nutrient Requirements of Horses*. National Academy Press, 5th Ed., Washington, DC, 1989.

NOTE: A BCS of 5 indicates the proper amount of dietary energy intake; 3 or less indicates inadequate energy intake; and 7 or greater indicates excess energy intake.

2-52. **Feed According to the Animal's Feeding Weight.** Dietary intake is best determined by adjusting the ration first to the animal's feeding weight (not necessarily current weight). Although accurate scales may not be readily available, the animal's weight can be reasonably estimated by measuring around the animal at the girth line. A tape measure is placed around the thorax just behind the withers and follows underneath the animal in the same manner as using a cinch strap. Measurements are taken following respiratory expiration. Personnel should hold the tape snug through several respiration cycles and note the smallest number.

2-53. This measurement is then converted to an estimated weight shown in Table 2-3. If the animal is not at BCS 5 at the time of measurement, the handler should add or subtract 200 pounds per BCS difference. For example, if the BCS was 3, he should add 400 pounds to the girth-measured estimated weight to arrive at a new feeding weight. This extra feed is needed to increase the animal's body. If the BCS is 4, the handler adds 200 pounds. Additional rations may need to be added for compensatory weight gain.

Table 2-3. Estimating a Horse's Weight From Girth Measurements

Girth Length (in)	Weight (lb)
55	500
60	650
62	720
64	790
66	860
68	930
70	1,000
72	1,070
74	1,140
76	1,210
78	1,290
80	1,370

Source: *Care and Feeding of the Horse.* Lon D. Lewis. 2d Ed., Williams and Wilkins, 1996.

2-54. **Increase Feed for Work Over Maintenance Level.** Animals should be fed two rations—one for idle (not working) and one for working. The inactive ration is calculated to maintain body condition but not to allow for increased weight. The idle ration can be predominantly forage. A small amount of grain is beneficial to ensure protein, amino acids, and trace mineral requirements are met. The idle ration should be of low enough energy content to preclude "typing-up" once the animal resumes strenuous work. It is best to feed fat, grain with high fat content, or soluble carbohydrates if they are accessible.

2-55. Working rations need to have increased amounts of high energy feed. The animals need this increase for the following reasons:

- Working animals have less time to forage and rest if working long hours. Even on optimal pastures, pack animals need 4 to 5 hours of grazing to meet their nutritional requirement.
- Working animals have an increased demand for energy.

- Grain and oil are easier to transport than forages.
- Small amounts of grain or oiled grain can easily be fed in nose bags (feed bags) during short halts.

2-56. Dietary components directly affect performance and onset of fatigue. High fat diets often enhance aerobic and anaerobic performance and delay fatigue. Fats or oil are preferred over either soluble carbohydrates or proteins. Likewise, soluble carbohydrates are preferred over protein. However, it is more difficult to balance the ration. Using regular grain is simpler, but not as beneficial. Table 2-4 shows amounts of forage and grain to feed for idle and working pack animals.

Table 2-4. Idle and Working Rations

Feeding Weight (lb)	Forage (lb)	Grain (lb)
800	14	7
1,000	15	10
1,200	16	13

Additional grain (or oil) increment for working is as follows:

One pound of high-quality grain per hour worked up to a maximum of 40 percent grain in the total ration. Over 40 percent will cause some animals to founder (lameness) or colic. Therefore, if more energy is needed, personnel should add oil to the grain and forage (for example, vegetable oil, corn oil, soy oil, or animal fat) in the diet. One pint of oil per 5 pounds of grain results in 20 percent added fat in the grain.

Example: An 800-pound pack animal working for 8 hours would need an additional 80 ounces (8 hours x 10 ounces) or 5 pounds.

NOTE: These are general guidelines that are suitable for most conditions. More precise rations can be formulated using the references cited at the end of this manual.

2-57. **Feed Animals as Individuals.** Some animals will require more feed for the same work even though they may appear to be of similar size, age, and temperament. These guidelines to feeding should only be considered as a starting point. Rations will have to be adjusted for the animal, weather, terrain, and altitude. In some cases as described below, the ration may need to be increased 50 percent or even more.

2-58. **Adjust as Needed to Maintain Ideal Body Weight.** Once the animal achieves a BCS of 5, the ration should be adjusted to maintain that BCS. Animals should be taped around the girth line regularly to track changes in body weight. Most people are too subjective and variable in their visual appraisals of weight and often will not detect weight change until the animal gains or loses 100 pounds. The ration will also have to be adjusted for work, weather, and altitude.

2-59. **Adjust Rations to Cold Weather, Hot Humid Weather, and Very High Altitude.** Horses and mules with a long, thick coat of hair can tolerate cold weather well if they remain dry and out of the wind. The energy requirement increases only about 1 percent for each degree Fahrenheit (F) below 18 degrees F. However, if the animals become wet and are subject to wind, the energy requirement can increase 20 percent.

2-60. Hot humid weather can decrease the animal's appetite. Therefore, it may be necessary to increase the palatability of the grain mix by adding molasses or other sweet-tasting feed. It may also be necessary to feed a high-grain and low-forage diet. Very high altitude can also decrease the animal's appetite. Likewise, it may be necessary to increase palatability of the ration.

2-61. **Change Forages Slowly.** The handler should change forages gradually over 5 days. If introducing animals to a new pasture, the handler should limit grazing to 1 hour the first day and increase by 1 hour for the following 4 days. The animal should be fed hay before turning out to graze to preclude overeating on new lush pastures.

TYPES OF RATIONS

2-62. There are three basic types of rations: garrison, field, and emergency. All three can be altered in quantities and substance depending on conditioning and training taking place, type and health of the animals, season, and combat situation. Table 2-5, page 2-19, shows a recommended allowance for garrison and field rations. The types of horses discussed in Table 2-5 fall in the following categories: small horses are usually found in overseas areas; light horses weigh less than 1,150 pounds; and heavy horses weigh more than 1,150 pounds.

Garrison Rations

2-63. The detachment uses these provisions at permanent or semipermanent operational bases where the pack animals are fairly idle. These rations contain a standard feed allowance of approximately 10 pounds of grain (8 for mules), 14 pounds of hay, and 5 pounds of bedding. Again, this ration can be increased slightly depending on the animal's condition or the training taking place. Idle horses can often be maintained on good quality hay or pasture without grain supplements.

Field Rations

2-64. While the detachment is deployed, the animals receive these rations so they can maintain condition and strength during heavy work. The field rations contain an allowance of about 12 pounds of grain (10 for mules), 14 pounds of hay, and no bedding. Such quantities and combinations of feed could cause a logistic problem in a combat situation and may be altered. If the situation permits, consider pre-positioning or caching the feed. Another way to prevent having to carry all the feed is to consider aerial delivery.

Emergency Rations

2-65. The detachment uses these temporary rations for a short time when the combat situation or environment prohibits the use of field rations. An emergency ration is a modification of a field ration for reasons such as logistic problems or the lack of forage in the operational area. This ration can vary greatly depending on the situation.

Table 2-5. Recommended Ration Allowances

		Grain (lb)	Hay (lb)	Bedding (straw/sawdust) (lb)
Garrison Rations				
Horses:	Small	7	14	5
	Light	10	14	5
	Heavy	12.5	15	5
Mules		8	14	5
Field Rations				
Horses:	Small	9	14	None
	Light	12	14	None
	Heavy	14	16	None
Mules		10	14	None

NOTES:
1. Bran may be substituted in amounts not to exceed 3 pounds for a like amount of grain. One-half pound of linseed meal may be substituted for 1 pound of grain.
2. The substitution of barley, rice, copra meal, or any other local product can be made for the grain ration.
3. The substitution of native grasses, bamboo shoots, or banana stalks can be made for the hay ration.
4. Fifteen pounds of corn fodder or grain sorghum is considered the equivalent of 10 pounds of hay.

FEED COMPONENTS

2-66. Types of grains and hay or combinations of available grains and hay will depend on the geographic location of the detachment. It is important, however, to come as close as possible to meeting all the nutritional requirements of the pack animals. Some of the grains, hay, and other items that may constitute a ration are described below.

2-67. Oats are the safest and most commonly used of all grains for the pack animal. Usually all other grains are combined with oats or regarded as substitutes. Oats may be safely fed in quantities up to 10 pounds per day but no more than 6 pounds when the animals are idle. Oats may be fed whole or crushed; however, crushing ensures more thorough chewing and digestion. Oats can be steamed or boiled for ill animals but new oats should not be fed until a month after thrashing.

2-68. Corn is best combined with oats and hay for feeding during the colder months since it has a tendency to produce heat and fat. When feeding ear corn, 6 to 12 ears are recommended depending on the amount of work being performed and the individual animal. If substituting corn for oats, the handler should make the change gradually because corn is considered a "hot" feed. That means it contains greater than 16 percent protein. If the change is made too quickly or the animal is fed too much, he could develop colic or founder. The handler can substitute about 2 pounds of corn for an equal quantity of oats weekly.

2-69. Barley is used extensively in Asia, Southeast Asia, and parts of Europe. It is considered a good grain and may be safely substituted for oats and fed in the same quantities. The change from oats to barley should be made gradually over an extended period, substituting 2 pounds of barley for an equal amount of oats weekly. Barley is very hard and should be crushed or soaked in water for 2 to 3 hours before feeding, but it may be fed whole. Barley is also a "hot" feed and the same care should be taken as with corn.

2-70. Rye is not regarded as a very good feed for horses and mules. If other feeds are scarce, it may be mixed with other feed such as oats or bran when necessary. Rye is very hard and should be rolled or crushed before feeding.

2-71. Wheat alone is not a safe feed for horses. It should be rolled and combined with a bulky grain or mixed with chaff or hay before feeding. One or two pounds should be given at first and the amount gradually increased to a maximum of 6 pounds per day.

2-72. Bran is the seed husk of grains such as wheat, rye, and oats, separated from the flour by sifting. It is an excellent food for the pack animal. Bran, having a mild laxative effect, is most useful as a supplement to a ration consisting largely of grains. It helps in building bone and muscle, has no tendency to fatten, and adds to the general tone and condition of animals. To supply the desired laxative and tonic effect, the handler can add necessary bulk and stimulate more thorough chewing. He should feed about 2 pounds of dry bran mixed with oats or other grain every day.

2-73. Rice—that is, rough rice, when rolled, crushed, or coarsely ground—may be fed in quantities up to one-half the grain ration. In an emergency, it may be fed in quantities up to 8 pounds daily.

2-74. Grain sorghum has a general food value of slightly less than that of corn. The handler should feed the pack animal the same amount of grain sorghum as he would feed corn, and under the same circumstances. Grain sorghum is less fattening than corn and has higher protein content.

2-75. Salt is essential to the health and well-being of all animals. The pack animal's need for salt is greatly influenced by the amount and type of work he is performing, since a considerable amount is excreted in his sweat. A supply of salt, adequate to replace that lost through sweating, is an important factor in preventing heat exhaustion during hot weather. Salt (8 to 10 ounces per day) should be available free-choice in a salt block. The handler should add no more than 4 ounces to the daily grain ration. If animals sweat excessively, they will need up to 12 ounces per day. Prolonged sweating depletes the body's store of potassium, which causes muscle weakness and early onset of fatigue. Potassium can be replaced with "lite salt" that is one-half sodium chloride and one-half potassium chloride. The handler can add 3 ounces lite salt per day in grain and add molasses to increase palatability and bind the salt to the grain. Otherwise, the salt will settle out of the grain as the animal eats. If salt blocks are used, they should contain trace minerals. Trace minerals are vitally important. Availability in forages and grains varies by region. Animals must receive supplements of trace minerals unless the rations are analyzed and guaranteed to be adequate. Trace minerals can be conveniently supplemented by providing trace-mineralized salt, either as a loose form or in block. The trace mineral mix should include zinc, manganese, iron, copper, cobalt, and iodine as a minimum. If the region is deficient or marginally variable in selenium, then it also should be added. Areas in which forages contain excessive selenium (over 5 parts per million of the total ration), including selenium in trace minerals, will cause both selenium poisoning and impaired use of other trace minerals. Selenium poisoning causes abnormal hair and hoof growth. The availability of a salt block eliminates the requirement to add salt to the feed except when lite salt needs to be used to counter potassium depletion.

2-76. Hay is the basic element of the ration that provides the bulk necessary for the proper performance of the digestive system. The pack animal should not be deprived of hay or something with similar bulk, such as straw, for any considerable length of time. Animals will suffer more on a ration of grain than on one of hay alone. Should the supply of hay normally required for the daily rations be diminished, the animals should be grazed or fed such roughage as can be gathered or produced locally. Oat straw is one of the best substitutes for hay if the oat heads are still on it. Otherwise, it is not a good substitute. Any straw, not spoiled, may be used, but barley and rye straw are not recommended. The handler should **never** feed moldy hay to horses and mules; cows can eat it without suffering ill effects, but horses and mules cannot. If bulky feed is unavailable, he can give the animals green or dried weeds or leaves as substitute roughage. All hays, except for the legumes, have nearly equal feeding values. Some of the more common hays are alfalfa, timothy, oat hay, and prairie grass. Alfalfa is protein-rich roughage of high nutritive value that more closely approximates that of grains than the common roughage. Alfalfa is an excellent source of calcium and vitamins. Being high in protein, alfalfa combines well with corn to create a laxative effect. If changing to an alfalfa ration, the handler should give about 2 or 3 pounds daily. The ration should compose not more than one-half of the hay allowance.

2-77. Timothy is usually considered the standard hay, although it is not particularly rich in digestible nutrients. Timothy mixed with clover gives a higher nutritive value and a better supply and balance of minerals and vitamins. However, the clover content should not exceed 50 percent. This mixture is common in areas where timothy is available, as timothy and clover are frequently seeded together.

2-78. Prairie hay, or wild hay, is produced from the natural grass growing on prairie land. Upland prairie hay, its feeding value being slightly higher than timothy, makes excellent hay when properly cured. Midland prairie hay, which is produced from coarser wild grasses growing on low land, is of lower feeding value and is not considered desirable.

2-79. Oat hay is oat straw before removing the oat grain from the head of the plant. It is cured in the same manner as other hay. It has a nutrient value about equal to that of timothy but is richer in protein.

GRAZING

2-80. Grazing alone can maintain idle animals satisfactorily. It provides good feed and exercise for the animals. As with any change in feed components, grazing periods should increase gradually over 5 days to condition the animal's digestive system. Grazing is an important source of roughage and should be used at every opportunity to reduce the consumption of hay, which may be difficult to obtain or be of questionable quality. Grazing also allows the detachment to carry less. Grazing on wet or frosted alfalfa or clover should be avoided to prevent gas colic. Except in an emergency, Johnson grass and grain sorghum should not be grazed.

PELLET FEED

2-81. This type of feed has several advantages over conventional hay or grain rations. Storage requirements are decreased, it is easily deliverable by air, and nutritious by-products can be included. Total feed intake is usually increased when pellets are fed. One disadvantage is that the cost is higher than conventional feeding. The contents of pelletal feed should be selected according to the type of roughage being fed. Alfalfa, for example, provides more protein and calcium than grasses such as timothy and orchard. Consequently, the protein and calcium content of the pellet should be lower.

WATER

2-82. Drinking water as a feed component is of utmost importance. The nutrients of the feed must be in a solution before they can be absorbed. During work, sweating and other physiological functions greatly deplete the water content of the body's tissues. To compensate for this loss, the body draws from the digestive tract. A deficiency of water in the digestive tract not only affects digestion but also may affect the general health of the animal by causing such problems as impaction colic and debility. An animal can survive for a considerable time without food but succumbs in a few days if deprived of water. Table 2-6 contains the chart for watering working animals.

Table 2-6. Average Daily Water Requirements for Working Animals

Species	Liters Per Day
Horse	30–50
Mule	15–30
Donkey	10–20
Buffalo	45
Ox	20–35
Camel, Dromedary and Bactrian	20–30
Llama	4
Caribou (Domesticated Reindeer)	Not Established
Dog	4
Goat	4
Elephant	169

FEEDING IN GARRISON

2-83. The times of feeding and watering while in the operational base or rear area should be fixed and regular. Feeding and watering, both as to time and amount, are based on the training, conditioning, and work being performed.

WATERING THE ANIMALS

2-84. It is important to offer the animals water and give them plenty of time to drink before feeding time. Normally, a pack animal requires about 8 gallons of water per day. However, the temperature and amount of work being performed will determine water requirements (Table 2-7). The animals should be watered three times per day under normal conditions and four times per day when operating in an extremely warm climate. Under ideal conditions, water should be available to the animals at all times when they are not being used. Feed should not be distributed while the animals are being watered because they will not water properly when they have feed available.

Table 2-7. Heat Conditions and Countermeasures (Increasing Severity)

Heat Condition Color Code	WBGT Index Degrees F	Water Intake Liters	Work/Rest Cycle Minutes Per Hour
Green	82–84	0.5	50/10
Yellow	85–87	1.0–1.5	45/15
Red	88–89	1.5–2.0	30/30
Black	90 and Up	2.5 and Up	Stop Work

QUANTITY AND FREQUENCY OF FEEDING

2-85. The stomach of the horse and mule is small and is unable to function properly while holding large quantities of food. Once the stomach is two-thirds full, the feed will pass through the stomach at the rate it is taken into the mouth. Therefore, the stomach functions properly when it is two-thirds full. If the animal is fed too much at one time, the stomach may become excessively distended and feed will be wasted by not being properly digested. A 3- to 4-pound feed of grain represents the approximate amount an average animal should be fed at a single meal. This feeding may be followed by a long and slow consumption of hay. Under these conditions the gradual passage of food into the intestines then takes place under favorable conditions. If the total amount of grain is increased, it is better to increase the number of feeds rather than to increase the size of the ration at each meal.

WORKING AFTER FEEDING

2-86. Working an animal hard after a full feed interferes with its ability to work and with its ability to digest the feed properly. The animal's ability to work is hampered by difficulty in breathing, which is caused by the swelling of the stomach and bowels against the diaphragm and lungs. Digestion is also accompanied by an increased flow of blood and secretions and increased muscular activity in the bowels. Hard work diverts blood to other parts of the body, tires the intestinal muscles, and reduces secretions needed to aid digestion. As a result, the animal suffers a loss of nourishment from the feed, may develop serious disorders of the digestive tract, and may die. The animals may be worked safely 1 hour after feeding.

FEEDING AFTER WORKING

2-87. The digestive organs of a tired animal are just as fatigued as the rest of the body. Therefore, a tired animal should not be fed a full ration. Most of the blood supply is still in the muscles, the muscles of the digestive tract are tired, and the glands used to secrete digestive fluids are not ready to function properly. The animal must be cooled and rested before feeding. The handler should give the animal small amounts of water at frequent intervals and permit him to eat long hay. After about 30 minutes of rest, he can give the animal a small portion of grain followed by the balance, a little at a time, after an hour or more. Failure to take these precautions frequently results in colic, laminitis, or both. The method of feeding just mentioned is time-consuming. If it is not possible to use this method, the handler should feed the animals once after waiting 1 to 2 hours after work.

FEEDING PROCEDURES

2-88. Feeding a small amount of hay before feeding grain stimulates an increased flow of saliva and gastric juices, takes the edge off the appetite, and quiets the nervous animal. In the morning, it is not necessary to feed hay first because the animal has probably been eating hay all night. An ideal way to feed hay is to keep it before the animal continuously by replenishing the supply frequently with small quantities. Feeding chaff with the grain adds bulk and forces the animal to eat more slowly, ensuring more thorough mastication.

FEEDING IN THE FIELD

2-89. Feeding pack animals away from garrison in a field or combat environment presents problems that are not present in garrison. The greatest problem is setting a regular schedule of feeding times. The hours that animals work in the field or under combat conditions are seldom the same every day. Therefore, to adhere to the principles outlined above, the animals must be fed smaller rations at more frequent intervals. It is very important to ensure every animal is fed a full ration every day to maintain the strength required to work in the field.

2-90. Due to logistic constraints, there may not be enough hay available to ensure the animal receives the required roughage. In this case, the animal must be allowed to graze at every opportunity to obtain sufficient roughage. Grazing should be allowed while at a bivouac location and at halts while on the march. A halt of an hour or more to feed grain should be planned if the duration of the movement will exceed 5 hours. Important points to consider when feeding in the field are covered below.

WATERING ON THE MARCH

2-91. Watering on the march should be done whenever possible, especially on hot days. When watering a string of animals on the move, such as at a stream crossing, the handler should allow the entire string to get in the water before letting any of the animals drink. Otherwise, the lead animals will drink and then try to move down the trail before the rest of the animals in the string have a chance to drink. The handler must watch to ensure all animals have a chance to drink before moving the string. He should angle the string upstream so the animals ahead do not foul the water. The handler should give the animals ample time to drink their fill and not be led away the first time they raise their heads. After watering, he should keep the animals at a walk for 10 to 15 minutes before increasing the gait or coming to a long halt. This action will prevent digestive disturbances.

FEEDING HALTS

2-92. The handler should try to plan the place for a feeding halt 2 or 3 miles past a watering point. He should give the animals a little hay after arriving at the feeding point and tying them to the high line. This procedure will help relax the animals and start the secretion of the gastric fluids.

FEEDING AT BIVOUAC

2-93. When it is necessary to feed and water at a bivouac location, the handler should wait at least 45 minutes after arrival to water the animals. He removes the bits if a full watering will be allowed. When watering at the bivouac location, the handler should lead the animals to the water on foot. He should lead no more than two animals at a time. When they drink, he should stand between the animals so they do not crowd each other. As stated before, the animals should receive ample time to drink and should not be led away from the water the first time they raise their heads. The animals should go to the water and leave the water together. Handlers should closely watch that the animals do not start pawing the water or lie down in it after being watered, as they often do. These actions will stir sediment on the bottom and make the water unfit to drink by other animals.

FEEDING OFF THE GROUND

2-94. Whenever possible, the handler should keep animals from feeding off the ground. As they eat off the ground, they pick up dirt and sand. The sand or dirt will accumulate in the colon. With time the animals will develop colic due to obstruction or they will develop diarrhea due to irritation. Feeding off the ground also causes premature wear of their teeth. For animals that are predisposed to forming sharp points and hooks, this will accelerate their development and subsequently degrade the animal's ability to eat.

WATER SHORTAGES

2-95. When water is scarce, its consumption will have to be regulated. If the bits are removed, animals can drink from a very shallow container. A small quantity will let an animal keep moving if he is given the water by the swallow instead of allowing him to take one long draft.

FEEDING HAY

2-96. When the animals are on a high line, the handler should break the hay from the bales and distribute it along the high line. The hay should be given in small quantities and replenished frequently. This procedure is especially important in damp climates or while it is raining. Moldy hay should never be given. As mentioned before, cattle can eat moldy hay without problems, but horses and mules cannot. The handler should break the bales of hay apart and distribute only as needed. Personnel should ensure the animals do not work the hay beneath them where they cannot get to it. The handler should place the hay flakes directly under the animal's head at the point where the animal's lead line is attached to the high line.

FEEDING GRAIN

2-97. When feeding grain in the field, the handler should feed the animals from a feed bag and ensure the feed bags fit properly. If they are too loose, the animals will toss their heads trying to get to the feed, and grain will be spilled and wasted. Personnel should watch the animals while they are feeding from a feed bag, and should never allow the animals access to water until they have finished eating and the feed bags are removed. While attempting to drink, the animals could get the feed bag filled with water and drown. Any leftover feed should be spread on a cloth to dry. The feed can then be used for the next feed. Grain should not be spread on the ground for the animals to eat. When the animals eat the grain, they also ingest dirt that can lead to colic. Any grain spilled on the ground in front of the animals should be swept. Figure 2-5 shows another method of feeding grain and commercial alfalfa cubes to the animal. If the handler is not using a feed bag, he can place about a 3-inch flake of hay on the ground with the grain placed on top of the flake. This method will keep the grain off the ground and keep the animals from ingesting dirt.

Figure 2-5. Grain on Flaked Hay

FEEDING ALFALFA CUBES

2-98. Foraged cubes are gaining popularity as an alternative to feeding long-stem hay (Figure 2-6). The cubes available may be 100-percent alfalfa, a mixture of alfalfa and grasses, or a mixture of alfalfa and whole corn plant. Alfalfa cubes can be used in feeding programs to replace a portion or all of the forage that animal handlers would feed their animals. They have high nutrient values for energy, protein, calcium, and vitamins. As with any feedstuff, there are advantages and disadvantages that the handler must consider when making the decision to use alfalfa cubes in his feeding program (Figure 2-7, page 2-28).

Figure 2-6. Commercial Alfalfa Cubes

CARE OF FORAGE

2-99. The care of forage is extremely important to the health of the herd. Feeding damp or moldy hay can cause colic and could disable a large portion of the herd at once. The handler should inspect the forage at the time of delivery to ensure the quality of it. In garrison, shelters are available to keep the forage dry. In the field, the unit should make every attempt to keep it dry. When the feed is packed on animals for transportation, it should be covered with a manta. This precaution will protect the feed from the elements and keep the animals from getting into it during the movement. During temporary storage, the unit should raise the forage off the ground by timbers or whatever else is available. This technique will keep the forage from getting wet and keep loose or stray animals from getting into it.

Advantages

- **Reduced feed waste.** Horses fed long-stem hay can separate the leaves from the stems and consume the parts they prefer; this does not happen with cubes.
- **Controlled feed intake.** It is easier for the animal handler to monitor and regulate the daily intake of cubed forage than long-stem hay.
- **Consistent nutrient content.** The nutrient levels found in cubes tend to be more consistent than hay.
- **Reduced dust.** Cubes have little dust and are therefore a good alternative to hay for animals with certain respiratory problems.
- **Reduced storage requirements.** Cubes can be mechanically handled in bulk and are denser than hay; therefore, they require less storage space.
- **Reduced transportation costs.** Cubes take up less space and are easy to transport on pack animals. The density of cubes allows trucks or planes to be loaded to their full legal capacity.

Disadvantages

- **Excessive feed intake.** The handler must feed cubes in a controlled manner to avoid overweight animals and, more importantly, to avoid serious digestive upsets.
- **Handling.** As with hay, alfalfa cubes require a storage area that provides protection from the weather to prevent spoilage caused by excessive moisture.
- **Cost.** Processing adds to the cost of the feed, and there may be additional costs associated with shipping. The major sources for cubed alfalfa are the western United States, western Canada, and Ontario.

Figure 2-7. Considerations for Using Alfalfa Cubes

Chapter 3

Animal Care and Training

The proper care and training of pack animals is essential to the health of the animals and their performance in the field. Without proper care, the animal's health and the unit's ability to complete its mission will suffer. Without proper training, the unit cannot rely upon the animal to behave in a manner that ensures mission accomplishment.

GROOMING

3-1. Grooming is essential to the general health, condition, and appearance of animals. It promotes good health practice and allows the handlers to bond with the animals. Animals in a herd bond by grooming each other; this action can be duplicated by the handler grooming the animal. In the wild, horses groom each other as well as rolling at will and rubbing against trees to maintain healthy skin. Domesticated horses must rely on humans to provide the opportunity for skin care. Grooming, no matter who does it (humans or horses), increases the circulation to the skin and releases the oils that provide luster to the animal's coat. It also provides an excellent time to inspect them for injuries. When grooming before movement, the handler should check the animals for injuries. He should also check the condition of past injuries, if any, at this time. When grooming after movement, the handler should check the animals for injuries sustained on the trail and for any evidence the saddle or harness may have chafed the animal. These precautions permit treatment of any problems before they get to the point of incapacitating the animal. The value of grooming depends on how thorough the handler does it. The animal handler obtains efficient grooming when he takes pride in the appearance of his animals.

WHEN TO GROOM

3-2. The animal handler should groom every animal thoroughly, at least once each day. He should always groom the animal before it leaves the stable area for work or exercise. Before saddling the animal, the handler should make sure the areas where the saddle pad, cinch, breast collar or strap, quarter straps, and britching ride are free of dirt and foreign objects. Failure to clean these areas will result in sores and could cause the animal to be unusable. On return from work or exercise, the handler should remove, clean, and put away any equipment. A heated, wet, or sweating animal should be cooled before grooming it. The handler should give the animal a brisk rubbing with a cloth to partially dry the coat, then blanket the animal (if a blanket is not available, he should leave the saddle pad on) and walk it until it is cool. The handler should check for injuries that may have occurred during movement. After removing the saddle and the animal is released, it will usually roll on the ground. The handler should hold off grooming until the animal has finished this process or all the effort that is put into grooming the animal will be a waste of time.

FM 3-05.213

EQUIPMENT

3-3. Each individual responsible for the care of animals should have a grooming kit. The basic kit consists of a currycomb, horse or body brush, hoof pick, and a grooming cloth (Figure 3-1).

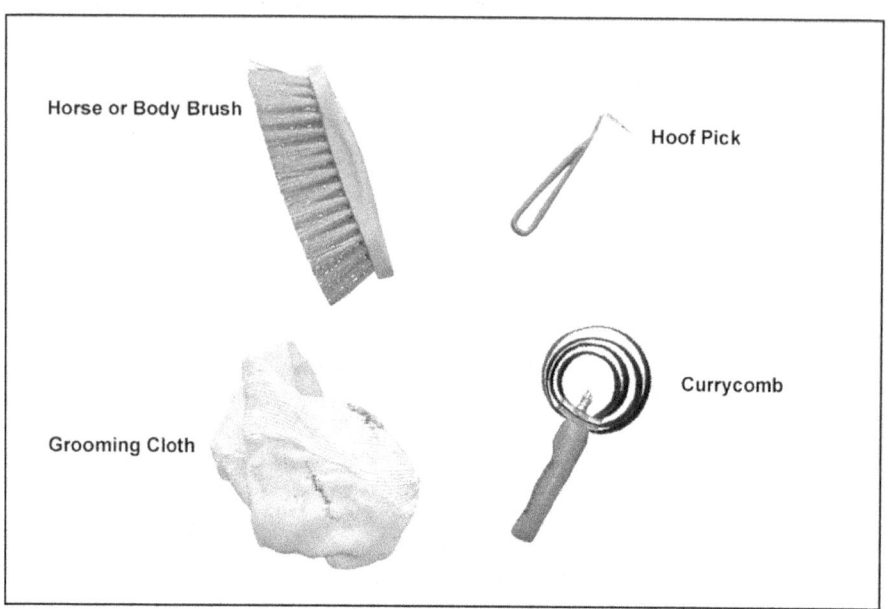

Figure 3-1. Grooming Equipment

Currycomb

3-4. The handler can use this circular metal device with sawtooth-like edges to remove caked mud, to loosen matted or dried skin and dirt in the hair, and to clean the body brush while grooming. He should never use it on the legs below the knees or hocks and never use it about the head. These areas are mostly skin and bone and the animal could be injured.

Horse or Body Brush

3-5. This brush is the main tool used in grooming. When used properly, the bristles of the brush penetrate through the hair of the coat and remove dirt and flaked skin from the hide as well as from the hair. Various brushes are available. Generally, it is better to use a stiffer bristle to better penetrate the hair on the coat.

Hoof Pick

3-6. Hooves should be cleaned daily and trimmed or reshod every 4 to 6 weeks. This small metal pick allows the handler to clean rocks and packed dirt from the hooves. The handler should always use the hoof pick in a downward motion toward the toe.

Grooming Cloth

3-7. The handler uses the grooming cloth to clean the body orifices and to give a final polish to the coat. He can make a grooming cloth, about 2-feet square, from old toweling or any other type of soft cloth.

GROOMING SEQUENCE

3-8. A prescribed sequence of grooming will enable the animal handler to groom effectively and thoroughly. The following paragraphs explain the recommended sequence.

3-9. The handler should first check the animal for any signs of lameness or abnormalities as the animal walks to the grooming area. If any exist, he should thoroughly clean the hoof of the affected leg with the hoof pick and look for rocks or other debris causing the problem. The handler should clean **each hoof** every day in working animals. One school of thought is that the soft dirt and debris packed in the hoof gives the animal extra cushion to walk on in rough terrain. Whether cleaned daily or not, the handler should check the animal's feet for thrush, torn frog, loose or missing shoes, and cracks. He should report defects at once to the medic or veterinarian.

3-10. The handler should use the currycomb in the right hand and the brush in the left. Beginning at the neck, he should brush over the left side of the animal with the currycomb. The handler continues down to the breast, withers, shoulders, foreleg, and knee; then smoothly transitions to the back, side, belly, croup, and hind leg to the hock. He should strike the currycomb frequently against his heel or side of the brush to free the accumulated dirt and dried skin. The handler should make sure the currycomb follows the natural lay of the hair.

NOTE: The handler should never use the currycomb on the animal's face or below the knees; these areas are tender and are made up of mostly bone.

3-11. The handler should brush the entire left side of the animal in the same order as above except brush the legs down to the hoofs. After a few strokes, he should clean the brush with the currycomb. The handler should stand well away from the animal when using the brush, keeping his arm stiff, and throwing the weight of his body behind the brush. A twist of the wrist at the end of each stroke will flick the dirt away from the hair. This use of the brush is not necessarily a separate activity from currying; the handler can do both at the same time with a brush stroke following each currycomb stroke. He should pass to the right side of the animal, change hands with the brush and currycomb, and groom the right side in the same order as above.

3-12. The head, mane, and tail are the last grooming features. In cleaning the mane and tail, the handler should begin brushing at the ends of the hair and gradually to the roots, separating the locks with his fingers to remove dried skin and dirt. He then wipes the eyes, nostrils, and lips and rubs the head, ears, and muzzle with the grooming cloth. The handler cleans the dock (fleshy part of the tail) and gives a final polish to the coat.

NOTE: To prevent the spread of skin diseases, the handler should wash grooming equipment and drying cloths with soap and water once a week.

ANIMAL INSPECTION

3-13. The handlers should inspect the animals, as indicated above, while grooming them. Good grooming offers the opportunity for close examination of the animal and the discovery of injuries or defects that otherwise might pass unnoticed. Correcting or treating these defects or injuries greatly reduces the number of noneffective animals in a unit. Along with wounds or other injuries present, there are others that are not immediately visible to the naked eye. For example, the animal handler can detect some joint and ligament injuries by inflammation or swelling. These injuries are often very slight and subtle, and the handler can best detect them by knowing precisely what is normal. The handler can establish an experience base by running his hands along the animal's body **daily** paying particular attention to the legs. If the animal flinches when touched in the area just to the rear of where the saddle rides, it is a sign that the animal's kidneys are sore. This pain might result from poor saddle placement or improper riding position.

GROOMING SICK ANIMALS

3-14. Handlers should never groom animals that are sick, weak, or depressed. These animals should be hand-rubbed at least once a day. The handler should wipe the animal's eyes and nostrils out with a damp sponge or soft cloth and clean the feet. This cloth should be left near that animal, not placed near other animals' equipment or feed, and **not** used on any other animal. The handler should groom animals with minor ailments in the usual manner. He should never clean or disturb animals with tetanus in any manner at all.

> **WARNING**
>
> Under no circumstances should an untrained Soldier attempt to shoe an animal. Improper shoeing or quicking of an animal will result in long-term damage to the animal and Soldier.

FARRIER SCIENCE

3-15. While not farriers, Soldiers must be able to replace, at least, a loose or missing shoe when a farrier is not available. The usefulness of a pack animal depends on the health and condition of its feet. The use of a "hoof boot" should be the first course of action when an animal loses a shoe. This item is part of the pack animal first-aid kit. The feet of a normal animal, due to their structure, require very little care or protection while on free pasture or even under light working conditions. At moderate work levels on good footing, an animal may require no more than cleaning and periodic rasping to trim and level its feet. The hind feet need only moderate care since they receive less shock. The front feet carry 60 to 65 percent of the load. As the workload increases or the terrain becomes more difficult, the animal's feet require additional care and protection. Shoeing protects the feet from excessive wear and enhances balance, support, and traction.

BASIC SHOEING

3-16. Shoes should be selected to suit the activity of the horse. Normally, shoes are for protecting the feet and supporting the limbs. The most common error made by the inexperienced horseshoer is to use shoes that are too small. The shoe should be as light as practical, but wide enough to offer sufficient protection to the bottom of the foot. The fit should be exactly to the perimeter of the foot at the toe and quarters. The shoe should fit slightly wider than the foot near the heels (approximately 1-1/2 inches of the heel of the shoe). This expansion can range from 1/16 to 3/16 inch.

3-17. Hoof growth, shoe wear, and the work required of the animal govern the frequency of shoeing. On the average, shoes may remain on the animal without change or adjustment for 1 to 2 months, though 3 to 4 months wear is occasionally possible. A farrier does the routine refitting. However, all handlers should have a basic understanding of the shoeing process. Several personnel in each unit should be able to replace or refit a lost or outgrown shoe.

FARRIER TOOLS

3-18. The farrier's kit should contain pincers, a pritchel, clinch cutter, hoof knife, hoof nipper, hoof rasp, blacksmith and driving hammers, clincher, fencing pliers, and assorted shoes and nails (Figure 3-2, pages 3-5 and 3-6).

Pincers 	Used to remove the shoe from the hoof, cut off excess length of clinches, or remove improperly driven nails.
Pritchel 	Used to enlarge the nail holes on shoes or to assist in extracting reusable nails from shoes.
Clinch Cutter 	Used to cut or straighten the nail clinch before removing the shoe. The handler uses the blade end for that purpose. He can use the other end as a pritchel.
Hoof Knife 	Used for cutting excess horn from the sole of the hoof and for trimming the frog, if necessary. It is available in right- and left-handed models.
Hoof Nipper 	Used to remove the excess growth of wall from the hoof when preparing to replace or refit a shoe.

Figure 3-2. Farrier Tools

Hoof Rasp	Used to remove excess hoof wall and to level the bottom surface of the hoof. The rough rasp is a hoof rasp that has become dulled. The handler can use it to remove the burr under the clinches and to smooth the clinch after the shoe has been replaced. He can also use the rough rasp to file away the clinches when removing a shoe.
Blacksmith Hammer	Used to shape the shoes to the animal's feet.
Driving Hammer	Used for driving the nails that secure the shoe to the hoof and for forming the clinches.
Clincher	Useful in finishing the clinches, especially when working on young or lame horses that object to having their feet struck with the hammer.
Fencing Pliers	Have several uses in the kit. They can be used as a hammer or for cutting nails and many other functions in caring for animals' feet. Can also be used to repair saddles and harnesses.

Figure 3-2. Farrier Tools (Continued)

RAISING AND HOLDING THE FRONT AND HIND FEET

3-19. An individual should know how to properly raise and hold an animal's foot before doing any grooming or farrier work so he can control the animal and have both hands free to work. Working with the front and hind legs is slightly different.

3-20. On front feet, the shoe should at least be large enough to cover the buttress of the heels. On horses with underrun heels, the shoe may need to extend past the buttress of the heels as much as 1/2 inch. This extension provides additional support to the flexor tendons and suspensory ligament.

3-21. The animal's leg should be held at a comfortable height for the animal, not a comfortable height for the handler. The handler stands with his feet apart, bends his legs, and keeps his back straight. This posture will free the handler to work with both hands (Figure 3-3, page 3-7). While the handler is working on the animal's leg, he should be alert to things around him so that he can predict the animal's intentions. Unless suddenly startled, an animal is quite predictable about wanting its feet on the ground. When an animal is getting tired and wants the front leg down, it will jerk its leg a few times to test the handler's grasp. If the animal is going to hop on its other foot to get away, it will move forward slightly with both hind feet in preparation for this move.

FM 3-05.213

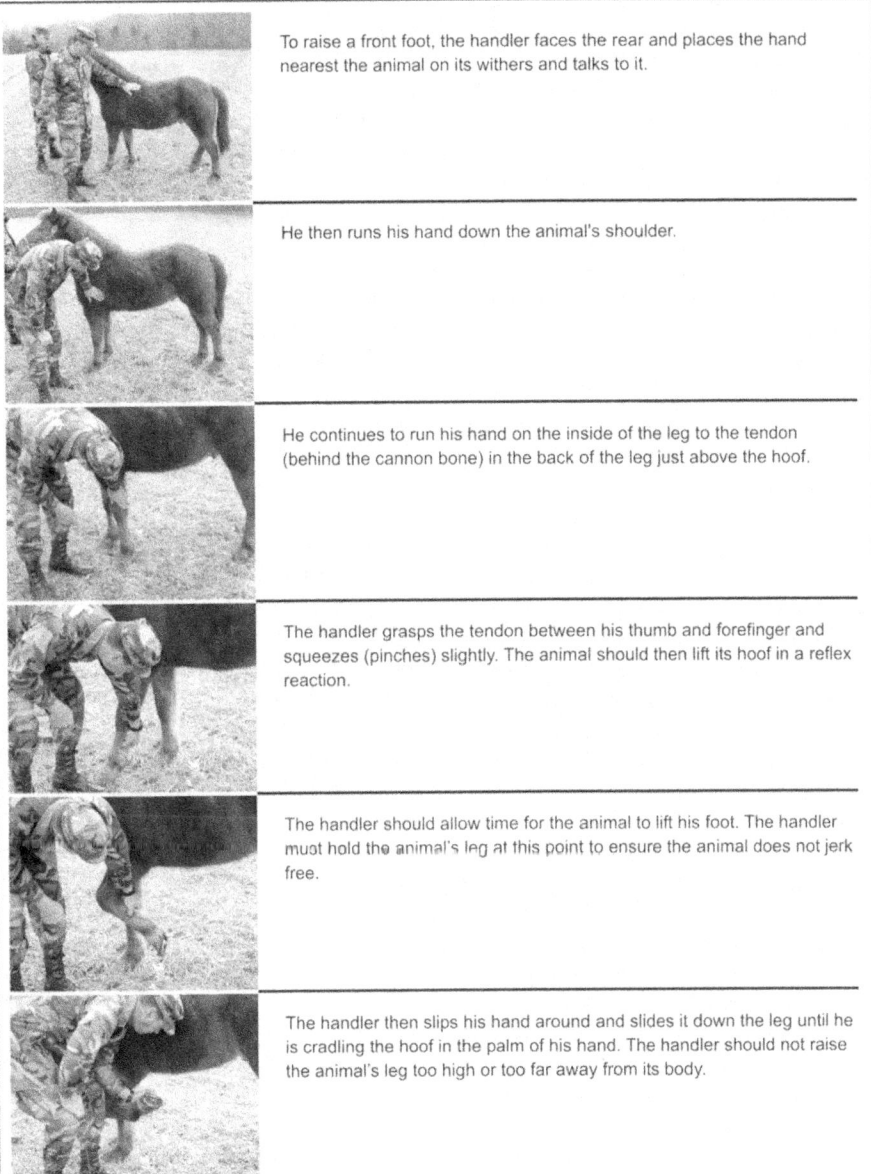

Figure 3-3. Raising the Front Foot

3-22. The farrier or groomer should not let an animal take its leg away every time it tries. The groomer should be considerate of the animal, but should not let the animal be the boss. If he does, the animal will soon get the idea that it can set its foot down whenever, which can be dangerous.

3-23. On most hind feet, the fit will be the same as the front feet except that it is desirable for the heels of the shoes to extend beyond the buttress of the heels 1/8 to 1/4 inch. This extension provides support and protection to the bulbs of the heels while the animal is stopping and turning. Figure 3-4, page 3-9, explains the procedures on raising the hind feet.

3-24. If the animal struggles and tries to kick, the handler should release his hold as the animal attempts to kick backward. By letting go, the handler will have time to move to the front and out to the side before the animal can pull back and kick him. If he releases as the animal is pulling the leg toward his head, the handler will not have enough time to move.

3-25. An animal will indicate when he wants his hind leg down. He will "cock" his leg by drawing it forward and upward toward his stomach, as if to "cow kick." If he does not cock his leg, he cannot kick with any force.

SHOE REMOVAL

3-26. To replace a shoe that has not come off the animal, the handler must follow a specific sequence to correctly remove the old or damaged shoe.

3-27. The handler examines the hoof approximately 1/2 to 5/8 inch from the shoe. He should see where the nails holding the shoe have come out of the hoof and are bent over or clinched. Using the clinch cutter and driving hammer, the handler cuts or straightens the clinches of the nails holding the shoe in place. He places the blade end of the clinch cutter under the nail point and taps the other end with the driving hammer. The handler can also use the rough rasp to accomplish the same thing by filing down the clinches. He should make sure the clinches are completely gone by straightening or filing. Failure to do so will hinder removal of the shoe or crack the outer hoof wall. This outcome will require a custom shoe with displaced nail holes. Driving a nail in the same location will likely "quick" the animal and make it lame.

3-28. The handler pulls the shoe by wedging the cutting edges of the pincers between the shoe and hoof at the heel and then exerts a quick thrust toward the toe. He repeats the procedure, alternating from one side of the shoe to the other, progressing toward the toe until the shoe is removed. Once the shoe is removed, he checks to see if all the nails came out with the shoe. If they did not, he checks the hoof to see if any nails remain. The handler removes any remaining nails with the pincers.

FM 3-05.213

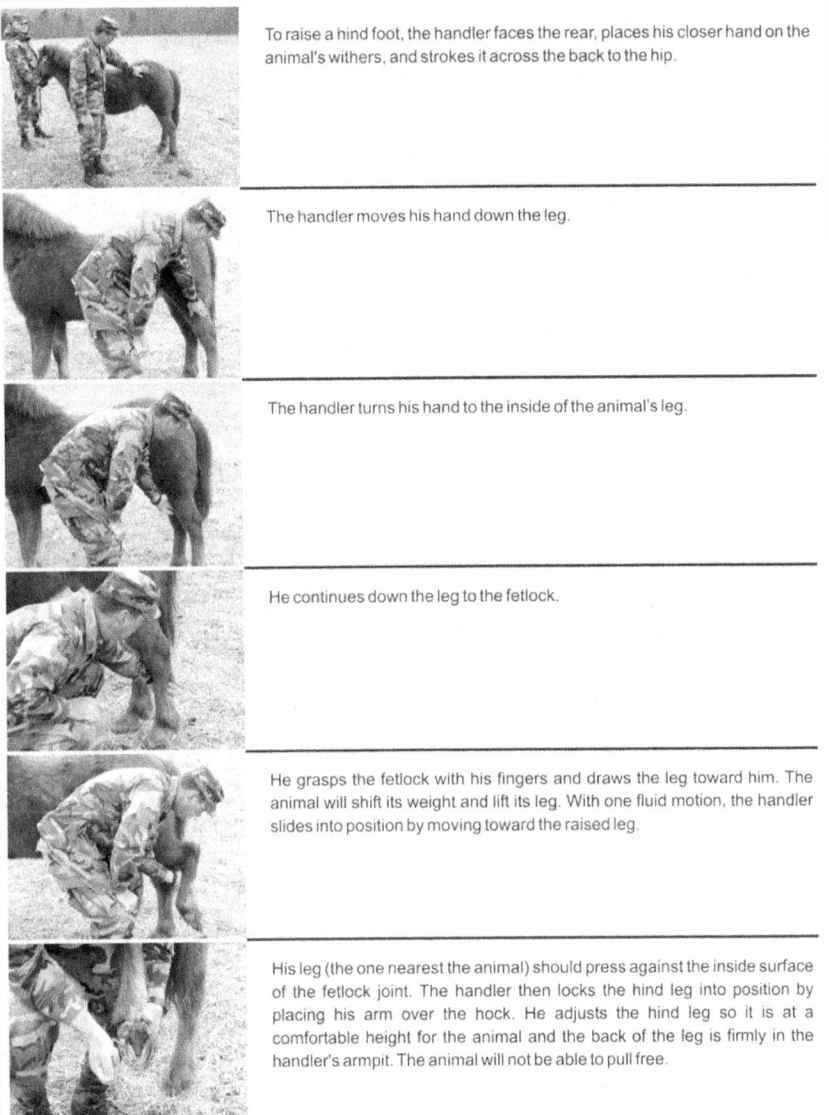

Figure 3-4. Raising the Hind Foot

HOOF PREPARATION

3-29. Using the hoof pick (contained in the grooming kit), the handler thoroughly cleans dirt and debris from the hoof. The same technique should be used on raising the front and hind foot. The handler then pares away the dead horn over the white line with the hoof knife. He should not touch the frog unless he cannot insert a hoof pick between the bars of the frog. In such a case, he narrows the frog slightly, but tries not to involve the surface touching the ground. The handler should be careful not to remove too much of the frog. The frog must be touching the ground and bearing weight. Lack of pressure on the frog will lead to contraction of the heels. The handler should level the foot with the rasp by using long, smooth strokes from heel to toe. He should be careful not to use too much rasp pressure when passing over the heel. He can always take off a little more heel later, but he cannot replace what has been removed. The handler should not expose live horn except over the white line and the outer border of the sole; the animal needs a covering of dead horn to protect the sensitive sole against bruising. He should also be sure to leave the bars intact as cutting away these structures can cause contraction of the heels and quarters. The handler then checks across the hoof to see that the surface is flat.

SOFT SHOES OR HARD SHOES

3-30. Regular metal horseshoes throw off a lot of sparks as the animals walk over rocky ground. These can easily be seen from a distance at night and may compromise location and activity. Additionally, these sparks can start fires if the tinder is very dry. Nonsparking horseshoes made of softer metal or hard plastic avoid these problems, but do wear out faster.

SHOE REPLACEMENT

3-31. The handler should fit the shoe to the hoof by holding it so that he can see the white line through the nail holes. He checks to make sure the shoe lies flat on the hoof surface without rocking. He eliminates any wobble between the shoe and hoof by rasping the high spot on the hoof. The handler may have to bend or shape the shoe as needed since it may have become deformed if it was worn loose for a time. An excess hoof wall in front of the toe of the shoe is not critical because it will be rasped away later in the shoeing process.

3-32. For most activities, six nails should be sufficient to attach the shoe to the foot. The handler may use eight nails on horses that travel in rough terrain or are troublesome about keeping shoes on. Height of the nails should be no lower than 3/4 inch above the shoes and no higher than 1 inch.

3-33. Before driving nails, the handler positions the shoe over the hoof and makes absolutely sure all nail holes are outside the white line. The white line measures approximately 1 millimeter wide and forms the juncture between the fleshy sensitive lamella (attached to the bone inside the hoof) and the hard insensitive lamella (attached to the hoof wall). Driving a nail into the white line can make the pack animal lame.

3-34. The handler secures the shoe to the hoof with nails and the driving hammer. He slightly bends the horseshoe nails near the point so they will turn out of the hoof when pounded into it. The other side of the nail is flat. The handler should always place the flat side of the nail so that it faces the outer edge of the hoof. He can be sure the nail is facing the right way by looking at its head; there is checkering on one side of the nail head. The checkering should always face the inside of the hoof. The handler grasps the nail between his thumb and forefinger and makes sure the flat side of the nail faces the outer surface of the hoof. He places the point of the first nail through the third nail hole on one side of the shoe, pointing parallel to the horn fibers of the hoof. He then taps the nail lightly several times to start it into the hoof wall.

3-35. To force the point of the nail through the outer surface of the wall at the desired spot, the handler continues to apply light blows until the nail is driven approximately 2/3 the required distance. He should then apply one sharp heavy blow on the head of the nail to force the point through the wall. The bevel on the point of the nail is effective only when driven rapidly through the horn. If the animal twitches or jerks away at any time during the nailing, the handler pulls the nail and checks for moisture or blood on it. The nail should exit approximately 3/4 of an inch above the ground surface of the foot to sufficiently hold the shoe. If there are old nail holes in the wall, the new nails should emerge at least 3/8 of an inch from the old holes. This length puts the new nails in hoof fibers strong enough to properly hold the nail. After the point emerges and the handler drives the nail head solidly into the crease of the shoe, he should immediately remove the point of the nail. The handler must do this quickly to avoid serious injury to himself should the animal decide to pull his leg away suddenly. To remove the point of the nail, he points the claws of the hammer toward the toe of the hoof. The handler places the claws onto the nail point as deep as possible and close to the hoof wall, bends the nail straight out from the wall, and rotates the hammer. This procedure will wring off the point and excess length of the nail. The handler then repeats the process with the same hole on the opposite side of the shoe.

3-36. After driving in two nails, the handler checks the shoe to ensure it still fits properly. He can adjust the fit by tapping lightly with the hammer to move it into place. He then puts the remaining nails in the shoe. It may be easier to start the first nail on each side of the shoe before driving either of them all of the way. The handler then sets the nails by striking the heads sharply with the driving hammer. He should make sure to hold a solid metal object (such as the flat side of the pincers) firmly against the end of the nail that protrudes through the hoof when he strikes. Some farriers think that the order in which clinches are set is very important. The recommended sequence is to set the two nails nearest the toe on the opposite side, then the quarter and heel nail on the first side, followed by the remaining heel nail. The handler cuts off the nail points with the pincers, leaving enough length for a proper clinch. The clinch should be approximately 1/8 of an inch long. A clinch that is too long can cause the wall to break should the shoe become caught on something and pulled off. The wall fibers of the hoof will break when the nail emerges. The handler removes the resulting burr under the seated clinch with the file side of the rough rasp. He should be careful to file away only the burr. The groove around the hoof should only be large enough to contain the clinch. The handler then uses the clincher to bend the clinches down into the groove.

3-37. Once again, the handler takes the pincers and the hammer to finish forming the clinch. He holds the pincer jaws against the head of the nail and strikes the end of the nail to seat it even with the wall surface. He should bend the nail abruptly at the hoof wall, ensuring the hammer travels parallel and as close to the hoof wall as possible. He can use the clincher for this step if one is available. The clinches are now finished. Using the file side of the rough rasp, the handler smoothes the nails leaving no sharp edges on the wall of the hoof. He then rasps and shapes the hoof to the shape of the shoe. Most of this rasping will take place in the area of the toe and anterior quarter. He must be careful not to rasp too much and rasp away the clinches. Clinches should be square in shape, embedded in the wall and smooth to the touch. The outer hoof wall should be smooth to the touch and free from coarse rasp marks. The periople at the hairline should be undisturbed.

FARRIER'S OBJECTIVE

3-38. The shoeing process will become clearer when it has been demonstrated. The intent of this section is only to enable the handler to replace a loose or missing shoe. If possible, he should fit each animal with an extra front and hind shoe. These shoes, their nails, extra nails, and farrier tools should be part of the standard equipment packed whenever the handler takes the animals out. The best way to prevent losing shoes is to catch loose ones early. The handler should check the shoes frequently by tapping them with a hoof pick. Shoes that are loosening will often sound different. The frequency to check them depends on the terrain. Loose shoes may not be lost from a pastured animal for several days. However, loose shoes in rocky terrain can be lost in a few hours. If a shoe becomes loose on the trail, the handler should remove it, if possible, to avoid its loss or injury to the animal. It would be ideal to replace lost or loosened shoes immediately, in garrison or on the trail. However, there will be times in the field when it is impossible to replace a shoe before 2 or more days have elapsed. This delay is typical and may have little or no adverse results. If this situation occurs, the handler can reduce the load of the pack animal or have a rider spend more time walking than riding his mount if it is a riding animal. Finally, there is the polyurethane boot that is tough, durable, and fits securely to the animal's foot. This boot slips over the hoof and clamps tight with a ski-boot buckle. These boots are easy to use and are remarkably durable. They are also a safe solution to the problem of mules who will not stand to be shoed. Animal handlers can use the easy boots (Figure 3-5, page 3-13) as replacements to shoes, to add traction or to soak an injured hoof. These boots are also used in correcting and treating hoof disease. These boots were designed to provide hoof protection for occasional and long-distance riding and to give traction on rocks, pavement, snow, and ice. The boot can be worn over iron shoes, in place of iron shoes, or only when needed. The adaptable inside strap can be adjusted, reversed, trimmed, or removed. To size, the handler must measure the animal's trimmed hoof across the bottom at the widest point. Easy boots can be ordered in the following shoe sizes: 00 (3 7/8 to 4 1/4), 0 (4 1/4 to 4 5/8), 1 (4 1/2 to 4 7/8), 2 (4 7/8 to 5 1/4), and 3 (5 1/4 to 5 3/4).

Figure 3-5. Easy Boots

FIELD TRAINING

3-39. A pack animal detachment will not always be able to work with animals that are trained and conditioned to pack. Even if the animals are trained to pack, they may need retraining or conditioning. In such cases, the handler should know some of the basics of training and conditioning animals. This section should provide enough information to evaluate the animals' current level of training and to train and condition them as necessary for pack operations.

TRAINERS

3-40. The attitude of personnel training pack animals is extremely important. A person assigned to train animals must have a better than average knowledge of animals. He must also have patience, tact, firmness, and a liking and aptitude for animal management. Experienced and knowledgeable trainers seldom need much in the way of restraint. They work confidently, orderly, and efficiently around the animals. As a result, the animals are cooperative, more productive, and sustain few injuries. Inexperienced trainers tend to use more restraint then necessary; are less confident and orderly; and consequently less efficient. Likewise, the animals buck more, are less productive, and sustain more injuries. A person who is afraid of animals or who will become frustrated easily with them will not do well. Above all, the trainer must not take out his frustrations on the animal by beating, kicking, or using excessive restraint on the animal. A good animal trainer combines an intelligent respect for animals with a lack of fear. An ideal pack animal trainer should be—

- Systematic.
- Patient.
- Tactful and resourceful.
- Moderate.
- Observant.

- Exacting.
- Logical.
- Tenacious.
- Consistent.

REQUIREMENTS

3-41. Previous experience and current abilities will determine the training needs of animals. Along with their physical conditioning, trainers must evaluate the animals on their level of proficiency in leading, packing, riding, picketing, standing, gaiting, swimming, balance, and conditioning to the sights, sounds, and smells of battle. Untrained animals never used for pack purposes will require a complete training and conditioning program.

EXERCISE

3-42. Exercise must be regular, graduated, and always within the capabilities of the animal. Working tired or unfit animals can cause accidents. If an animal is idle for a considerable time, the handler must repeat its process of conditioning. The minimum period of exercise necessary to maintain an animal in working condition is 2 hours daily. Animals need not be maintained in peak condition for field duty at all times. However, they should remain in such condition that a relatively short period of carefully scheduled work will put them in fit condition for anticipated duty. The kind and amount of exercise given to animals depend on the type of work they are to perform, their current condition, and the number of individuals available to exercise them. Although the most satisfactory exercise is gained by assigning one or two animals to an individual, pack units (with limited manpower) will normally have to adopt other methods. The trainer should begin any period of exercise with 10 to 15 minutes of walking to ensure good circulation, particularly in the feet. Instead of hard surface roads, the trainer should try to select dirt roads for exercise because they are easier on the animals' feet and legs. He can also work the animals on trails or cross-country to maintain their fit condition. If exercising two animals, he should try to always ride one out and the other in and lead each alternately on his right and left. This habit will get the animals used to being both led and ridden. The trainer should end each exercise period with a 10-minute walk to return the animals to the stables dry and breathing normally. The walk is the prime-conditioning gait. Walking develops muscle, trotting improves balance, and galloping develops wind. If the trainer uses either of the faster gaits in excess, the animal will lose rather than gain condition. He can determine the length of trot periods by the condition of the animals, but in no case exceed 7 minutes. The routes for exercise from day to day should vary so that the animals will not recognize the route and try to hurry back to the stables. Also, using different routes relieves monotony. The training and conditioning program should be set up according to the amount of time available and the level of proficiency of the animals and handlers.

CONDITIONING

3-43. An animal requires good conditioning to perform the work demanded of it without injury to its body and muscular organs. The handler can acquire and maintain good conditioning through a progressive program of proper exercise and feeding. Objectives of the program are endurance, stamina, a good state of flesh, and resistance to disease. The handler can attain these objectives only by proper feeding and long periods of conditioning that work at the slower gaits, mostly the walk. He must condition both pack and riding animals to carry the weights required in field operations.

3-44. Equine, as well as human, physiology naturally but gradually compensates and strengthens (shapes up) during times when more work is needed. But like humans, horses can quickly lose conditioning or "get out of shape" during long periods of rest. That is, a strenuous mountain ride would not be an appropriate activity just after 3 months of pasture or stall rest. These factors include cardiovascular fitness, respiratory fitness, thermoregulation, muscle fatigue, and skeletal fatigue.

Cardiovascular Fitness

3-45. The resting heart rate of horses is approximately 35 beats per minute (bpm) and can reach up to 250 bpm during extreme high-intensity exercise. Each beat can pump between 0.8 and 1.2 liters of blood. Therefore, a horse exercising at maximum intensity can pump enough blood to fill a 55-gallon drum in 1 minute! As a horse becomes more fit, the stroke volume increases permitting sufficient oxygen transport with fewer beats. Conditioning exercise will improve blood circulation through muscles. As blood circulates more efficiently through muscle, more oxygen is made available, and more heat can be removed.

Respiratory Fitness

3-46. Respiration is, of course, how oxygen is introduced to the horse's blood. Limiting factors can influence the amount of available oxygen. These include the volume of the lungs, the diameter of the airway from the nostrils through the windpipe, and their gait (since horses breathe in rhythm to their stride). One reason horses breathe faster during hard work is directly related to the pH of the blood. The more acidic the blood (from carbon dioxide and lactic acid), the harder the horse will breathe to get rid of excess carbon dioxide, as well as to take in sufficient amounts of oxygen.

Dissipation of Heat

3-47. Working muscles produce heat. Horses have two ways to remove it. One is through breathing heavily. Horse's lungs are very large and the expiration of hot air and inspiration of cool air help to reduce the temperature of the body, especially the area around the heart. The second is, of course, through sweating. As a horse overheats, blood vessels in the skin become dilated so that they can hold more blood. Then evaporation and transpiration of sweat helps to cool the horse much like a swamp cooler can cool a house. The sweating mechanism works best in cool, dry air. Warm or humid conditions may cause many horses to have more difficulty in keeping cool.

3-48. Fat horses and horses with heavy muscling are not able to eliminate heat as efficiently as leaner and lighter muscled animals. Safety becomes a concern when overheated horses become lethargic and uncoordinated. Conditioning exercise (particularly walking and trotting) in balance with proper feeding will remove body fat and improve the horse's ability to dissipate heat. Also, as blood circulation through muscles improves, the horse's heat can be more efficiently carried to the skin for cooling.

Muscle Fatigue

3-49. Working muscles need fuel. During normal (aerobic) exercise (walking or trotting on level ground), fuel stored in muscles is combined with oxygen from the blood to produce energy and motion.

3-50. During intense exercise (exercise that causes the heart rate to exceed 150 bpm), oxygen is depleted more quickly than it can be supplied. Many of the cells in the muscles then switch to an energy system that does not require oxygen. The main problem with this anaerobic system is that it requires over ten times the amount of fuel to produce the same amount of energy. This system also produces lactic acid as metabolic waste. If too much lactic acid accumulates in one area of the muscle, inflammation and soreness result. With a proper conditioning regimen, the horse will gradually improve his ability to both take up oxygen and deliver oxygen to working muscles. Conditioning will also improve his ability to rid muscle tissue of metabolic wastes before they can build up and cause any damage.

3-51. One good way to help ensure that the horse's muscles remain healthy after a high-intensity work effort is to allow 30 minutes or so of walking and light work to allow the horse to "cool out" before going back into a trailer or stall. During this cool out period, lightly active muscles allow blood and lymph fluids to circulate and rid muscular tissues of metabolic waste and heat much better than if the muscles were not moving.

Skeletal Fatigue

3-52. The skeletal system includes the horse's bones, joints, tendons, and ligaments. If overstressed, skeletal failure can cause abrupt and serious injury to both the horse and rider. Overworked horses are more likely to suffer sprains and strains when at a crucial moment in the horse's stride a particular muscle fails to contract, resulting in a momentary and sometimes repeated malpositioning of a related joint or ligaments.

3-53. During exercise, the horse's bones, joints, and ligaments are constantly changing to adapt and compensate for activity changes and to the rider's added weight. How they compensate is specific to the type of work performed. For example, roping, cutting, or barrel-racing horses will not necessarily be prepared for a strenuous day of climbing steep grades on a long mountain ride.

3-54. One big problem with skeletal conditioning is that compensatory changes occur much slower than circulatory, respiratory, and muscular conditioning. It takes approximately 60 days of 5-day/week riding for the density of a horse's cannon bones to adapt to more strenuous activity and to carrying the added weight of a rider. Therefore, while the horses may feel fit, the vast majority of personnel (weekend riders) do not ride enough to cause any significant changes in skeletal fitness. Unconditioned bones, joints, and ligaments are especially susceptible to shock, twists, and torsion. For this reason, riders should always be careful to slow horses to a walk on surfaces that are hard, slippery, uneven, or deep.

3-55. To sum up, before the handler can expect the horse to perform a new type of activity, he must consider how well the horse's varied systems have been prepared. To avoid problems, the handler should always introduce high-intensity work gradually, allow plenty of time after a high-intensity work effort for the horse to cool off before returning to a stall, and understand that physical limits vary with weather conditions or between horses and that fatigue may set in sooner than expected.

3-56. Factors that can cause lameness on the trail for an out-of-shape horse include the following:

- Work intensity exceeds oxygen supply to working muscles.
- Aerobic function is limited.
- Anaerobic function is increased.
- Rider fails to recognize symptoms of anaerobic onset.
- Lactic acid production increases.
- Circulation is inadequate to remove lactic acid.
- Fuel in individual muscles is depleted.
- Fatigue and soreness cause changes in movement.
- Changes in movement cause forces to be distributed differently on supportive ligaments, joints, and bones.
- A bad step causes a fall or injury.

3-57. Signs that indicate the horse is approaching his limitations are as follows:

- Panting or blowing respirations.
- Heart rate more than 150 bpm.
- Profuse sweating.

NOTE: If the above symptoms occur, the handler should do the following:

- Stop and rest.
- If the horse is still breathing heavily or the heart rate has not gone below 100 bpm after 5 minutes of rest, discontinue its exposure to intense work for a few days. For instance, if trail riding, it would be best to dismount and walk until the horse is rested, then choose less strenuous routes on subsequent rides until its conditioning improves.
- If the horse continues to blow and sweat after several minutes of rest, it may have been overworked. Immediately discontinue riding, cool out the horse, and consider it off limits for at least a week.

PROGRAM DEVELOPMENT

3-58. At the beginning of training, animals may be in poor physical condition and unaccustomed to hard work. To properly condition animals, yet avoid injury, the handler should ensure operations are long in duration but mild in character. According to the training principles mentioned above, each animal should receive advanced training that includes gentling, leading, riding, standing, packing, gaiting, swimming, and seasoning to battle conditions. The handler should use actual field movements, progressive in length, throughout the training period to build up endurance. He can conduct a part of such field movements and other phases of training at night to prepare the animals for night operations. The handler should pay close attention to the animals during night training to determine if any are night blind. If so, he can supplement their diet with Vitamin A. Personnel who train animals should know how to use restraining devices for controlling animals. Such devices include the twitch, cross tie, and blinds. Personnel should exercise great care when using such devices. It is best to use the mildest and least dangerous method of restraint necessary to achieve the desired results. Oftentimes, kindness, perseverance, and tact will accomplish the desired purpose without using restraints.

3-59. The 21-Day Pack Training Plan outlined in Appendix B is based on a system developed by the United States Army Artillery (Pack). This plan was developed for use with mules and may need to be adapted for use with other species. Because of the length of this training plan, time may not be available to conduct all of the training for the animals and ARSOF personnel, but the plan will provide guidelines and techniques for planning training.

3-60. During the training period, the animals have been made acquainted with many of the things to be encountered in open country, mountainous terrain, and jungles, including objects (tractors, newspapers, helmets, cans), strange noises (vehicle exhausts, gun shots, locomotive whistles), and odors (iodine, ether, smoke). These things should be presented in such a way that the animals do not associate them with harm or pain to themselves.

3-61. Obstacles should include loading ramps, steps cut in banks, small ditches, corduroy roads to cross swamps or bogs, and narrow bridges such as grease racks for cars and trucks. (The handler can simulate narrow bridges by constructing a walk, six-tenths of a meter wide and not over one-third of a meter high, over a pool of water; later, he can narrow the bridge and make it higher. He can also cut and place trees over a small stream, and with the top of the tree squared off, make a practical field bridge.) The handler should tie loads that rattle (a #10 can with stones in it) during the training. He should select narrow trails and make the animal do a considerable amount of climbing. He should have the animals go down slides, and walk up and down banks. The handler should load and unload the animals into trucks and, if appropriate, rail cars. He should always keep the animals coming along slowly and try to keep them in BCS 3.

RESTRAINING ANIMALS

3-62. Many ingenious devices have been developed over the years for restraining animals. Some have proven useful and humane and have helped to quiet and train animals. Others, although they could temporarily quiet and subdue an animal, make its attitude worse than it was originally. The handler should use restraints **only** when needed and use the least amount of restraint needed.

Blinding

3-63. This method is the easiest way to restrain an animal. The handler makes sure to tie the animal securely to a tree or post. Then he takes a piece of cloth (a gunnysack or jacket), or anything else that can be placed over the eyes, and ties it around the back of the animal's head. When blinded, an animal will seldom try to move. Once the animal responds to the restraint, the handler can vary the amount and use the least amount of restraint necessary. Animals should always be rewarded for good behavior. By responding with less restraint to the animal when he behaves correctly, the handler will gain his trust and respect. The animal in turn will be more productive and easier to manage.

NOTE: Under no circumstances should an animal be moved while it is blinded.

Twitches

3-64. An animal handler uses twitches because they are the handiest and most common method of restraint. Because the twitch shuts off circulation in the lip, the handler should never use it continuously for an extended time and never with greater force than is necessary. The handler makes a twitch by running a small piece of rope or chain through a hole in the end of a rounded piece of wood 2 to 5 feet long, such as a pick handle, and ties it in a short loop (Figure 3-6). He passes the loop of the twitch over the upper lip, which he seizes by hand and draws forward, taking care to turn the edges of the lip in to prevent injury to the mucous membranes. The handler tightens so the twisting points up over the outside of the upper lip. If the twitch is twisted down toward the end of the upper lip, it will easily slide off as the animal shakes its head or pulls away. He then tightens the loop by twisting the stick until he obtains sufficient pressure. Light changes in pressure with increases against resistance and decreases as the reward for obedience will help keep the animal's mind off the reason for the restraint and reduce the need for severity. These changes also help in training the animal to be more controllable. The handler should seldom use the twitch as a restraint while saddling or packing animals. He should use the twitch mostly when restraining for medical treatment or shoeing.

Figure 3-6. Twitch

> **CAUTION**
>
> Any long-handle twitch can be very dangerous. If the animal is successful in pulling free, the twitch becomes a high-velocity, high-energy object that can fly 50 meters or more. Fractious horses and mules can eventually pull these twitches free from even the strongest and most determined handler. Other twitches are safer. One is the pincher style that can be clipped to the halter.

3-65. A handler also uses a type of twitch made from a 16-inch piece of cord, such as parachute cord, and a horseshoe. He ties the cord with a square knot to form a continuous loop. He then twists it and folds to form two smaller double loops. He places the double loops over the upper lip in the same manner as a long-handle twitch. The handler inserts one-quarter of the horseshoe into the loop so that the cord will tighten against the side (quarter) of the horseshoe. He turns the horseshoe to tighten the loops in the same manner as tightening a tourniquet with a stick. The handler should tighten so the twisting points up over the outside of the upper lip. If the twitch is twisted down toward the end of the upper lip, it will easily slide off as the animal shakes its head. The handler then inserts the opposite quarter of the horseshoe into the ring on the halter. If this twitch is applied properly, it will not come off as the horse or mule pulls back and shakes its head. If the animal tries, the handler should let him. After a few failing attempts, the animal will cease fighting and the handler can proceed.

Cross Tie

3-66. Trainers often use the cross tie as a mild form of restraint. It consists of securing the animal's head in a normal raised position by two tie ropes extending from the ring in the halter to opposite sides of the stall or between two trees (Figure 3-7, page 3-21). When the cross tie is properly adjusted, the trainer may use it to his advantage while grooming, saddling, or doing any work around the animal. The cross tie also prevents an animal from chewing a wound and from lying down when he needs to be standing.

Distractions

3-67. Often a distraction rather than a restraint will enable the handler to accomplish what task needs to be completed. Three forms of distraction are as follows:

- Rubbing the lower eyelid.
- Tapping over the nasal sinuses.
- Hand twitches to the point of the shoulder.

3-68. The handler can distract the animal to complete short procedures merely by rubbing the animal's lower eyelid with his index finger. He places the palm of the hand on the side of the face so that the hand will move as the head moves. Failure to maintain contact with the palm against the face will often result in inadvertently touching the eye as the animal moves its head. The handler then rubs the lower eyelid with light to moderate pressure, but ensures the index finger does not slip into the eye. Lightly tapping over the nasal sinuses will also distract the animal enough to complete short procedures. The handler taps the fingers over the nasal sinuses with light to moderate pressure.

Figure 3-7. Cross Tie

Other Restraints

3-69. Trainers use additional types of restraints when treating an animal whose condition is such that complete immobilization of the part to be treated is required. They use these methods only under normal conditions. Giving tranquilizers makes the more severe restraints unnecessary and reduces the danger to the animal from abuse at the hands of the ignorant or inexperienced. Other restraints include the—

- Side stick.
- Muzzle.
- Knee strap.
- Casting rope.

- Side line.
- Running "W." (Both front legs are restricted by rope and can be pulled out from under the horse. This restraint should only be done on soft ground to prevent injuring the horse's knees.)

CALMING AN ANIMAL

3-70. Fear is one of the animal's strongest instincts. If fear is allowed to remain a dominant instinct, the animal cannot be trained satisfactorily to do the work demanded of it. The goal of the trainer throughout the training period should be to gain and maintain the confidence of the animal. Horses and mules have a remarkable memory and tend to remember the unpleasant experiences longer than the pleasant ones. Retraining thus becomes more difficult than initial training, especially with a young, impressionable animal. Rewards for accomplishment are extremely valuable in the gentling process. Patting the neck, rubbing the head, and hand-feeding are good aids in gaining the confidence of the animal. The unwarranted use of whips, switches, or other devices to inflict pain or restraint should not be allowed in training. A willing, confident animal will work to the extent of its physical ability, but the scared, reluctant animal will expend less productive energy than its trainer.

CATCHING AN ANIMAL

3-71. The most convenient method of catching loose stock is by just walking up to the animal either in a corral or in the field. However, the handler must first have gained its trust. Only through love and kindness to the animal will the handler gain trust and respect for his presence at the animal's neck. Once the handler starts to catch a loose animal, he should not give up just because the animal walks off or spooks. The handler should not leave to go to another animal. If the handler is not firm, he will soon have an animal that is impossible to catch without either roping or cornering it in a corral. The handler will have taught the animal to avoid him by letting go at the first sign of resistance. The handler should maintain eye-to-eye contact while he tries to catch a loose animal. The animal often communicates his intentions through his eyes. The handler should also watch the animal's ears. If they point backwards or cock to one side, he does not have the animal's complete attention, and he must have it to catch it. The animal must respect the handler above all animals in the immediate area. The head of an animal being caught is not always the first thing a person should touch. Often the middle of the neck or the back area is the best area. Rubbing the animal about the withers and back will enable the handler to step in close to it and then hook his arm over the animal's neck. Some animals will try to pull away at this point. If the animal pulls away from the handler more than once, he should slip the halter rope over the animal's neck when it starts to scratch or fondle him. The handler will then have a loop to hold it. When the animal starts to move away, the handler can stop it firmly. The animal should immediately be rewarded for his obedience with a few reassuring pats and caresses. The handler must never knock the animal around for responding to the halter rope around its neck. If he does, the next time he wants to catch this animal, he may not be able to get near it. Last, but not least, a little grain in a bucket or feed bag will usually let the handler catch most animals that are used to being grained. If the animals get used to being grained immediately after being caught, their anticipation of the grain will make them very easy to catch. Having an apple or carrot to give the animals after catching them is another way to ensure their cooperation, but the handler should be sure to always give it to them. The handler should never tease the animals with a treat just to coax them into getting close enough to catch, then not give it to them. As mentioned before, horses and mules—especially mules—have good memories and will not forget such an insult.

BREAKING TO PICKET

3-72. Picket breaking is an important thing to teach animals to ensure against halter pulling. If the handler picket-breaks an animal before attempting to saddle or pack it, the impression of secure fastening will never leave the animal. It will feel that any strain put on the halter or halter shank is useless. Trainers have used the following method for picket breaking successfully for many years. It saves time and labor, since the animal does most of the work, and seldom has to be repeated. The handler catches the animal and puts a strong halter with a large halter ring around its neck. He should use a good, strong, soft rope, 1 inch or more in diameter. The handler runs one end through the halter ring and ties a fairly snug bowline around the animal's neck. He ties the other end securely to the middle of a smoothly trimmed log approximately 12 inches in diameter and approximately 20 feet long. The handler then places the log in the middle of a smooth, open piece of ground where the animal will not be able to become entangled with anything except the above-mentioned equipment. If no logs are available, he can use something like a large truck or tractor tire. The animal may pull, tug, and jerk on the log and try everything it can think of to get away from the log but without success and without harm. Soon the animal will learn that it cannot go anywhere and that a halter and rope mean escape is impossible.

3-73. The handler should teach an animal to stand still when the lead rope is hanging. Trainers call this kind of restraint a "ground hitch." Breaking to a ground hitch will prevent an animal from straying or running away if it comes loose from the pack string. One way to train an animal to stand with the lead rope hanging down is to run the lead rope through a ring attached to a spike in the ground. The handler places a set of hobbles on the animal and runs the lead rope through the ring and then ties it to the hobbles. When the animal tries to move, he will not be able to. The animal soon learns to stand still when the lead rope is hanging. During the training period, the handler makes sure to observe the animal from a distance to ensure no harm comes to it.

LEADING

3-74. The handler must teach all animals to lead. He first teaches them to lead next to a dismounted packer. Then he teaches them to lead alongside well-broken animals. Leading should be at the walk, as daily exercise, until the new animals lead quietly and have improved sufficiently in condition to allow them to start their instruction under saddle.

RIDING

3-75. The handler should break all animals to riding and ride with regularity during training before initial work under the pack. Since the mouths of young or untrained animals are tender, the handler should not use bits during the initial riding periods. He can attach the reins to the halter or use a hackamore (a bridle with a loop capable of being tightened about the nose in place of a bit).

PACKING

3-76. After the animal has been ridden for about 10 days, the handler can mount a packsaddle on him. For the first few days, the animal's packsaddle should not be loaded. The handler should ensure the animal is thoroughly familiar with the method of saddling and unsaddling. He should also make sure the animal stands quietly while packers are working before he places any load on the saddle.

3-77. Training under load should be progressive. Initial loads should be light, single-side loads, such as sacks of oats. The handler should not attempt to condition the animal for top loads until it is completely conditioned for full side loads. When the animal is comfortable with this weight, the handler can add top loads to the side loads. He should gradually increase the load from day to day until the animal is carrying his full payload of 200 to 250 pounds. After the handler trains the animal to stand, he should train it to stand while being saddled and packed. If at first the animal does not stand quietly, the handler may need to use the blind. As mentioned before, the blind is an exceptional aid and should only be used when the animal clearly indicates the need. An animal should **never** be moved a single step while blinded.

GAITING

3-78. Animals individually led by dismounted drivers maintain a rate of march from 3 1/2 to 4 miles per hour and seldom move at a gait faster than a walk. Those led by mounted drivers may be required to take any gait demanded by the situation. The walk and amble are the most satisfactory gaits for mules. The animals can maintain the walk for long periods at 4 miles per hour (a natural gait for the animal) and tend to disturb the load less than at any other gait. The amble is an acquired gait, easily acquired by the mule, that increases the rate of march to slightly more than 5 miles per hour. This gait is also easy on the load and hitches and, due to the increased rate of march, is favored over the walk on fairly level ground. Trainers teach the amble to pack mules by increasing the rate of march gradually during the training period. In hurrying to keep up with the column, mules will first break into the amble for a few steps. These periods at the amble will gradually increase until they become confirmed in this gait.

SWIMMING

3-79. A handler must teach pack animals to swim boldly and freely. Although they are naturally good swimmers, some animals are initially afraid of the water and will resist entering it. When they do go into the water, such animals fight their environment and swim very poorly. The handler should introduce the animals to the water quietly, coax them to wade through shallow water at first, and then lead them into increasing depths gradually until they must swim. Known good swimmers should accompany the green swimmers during this phase of training to give confidence to the novices. Chapter 8 explains water crossing.

BATTLE INDOCTRINATION

3-80. The handler should try to mentally condition his pack animals to as many of the sights, sounds, and smells of combat as possible. Once the animals become accustomed to these sensations, he can feel assured about their docility and good conduct in the field. The handler must conduct this mental conditioning or battle indoctrination so that animals will not associate the sight, sounds, or smells with harm or pain to themselves. Following are some tips:

- Conduct a portion of the training in marching, packing, and unpacking close to motor parks. The animal will then be subjected to the—
  - Sounds of motors being started and warmed up.
  - Pounding of metal on metal and an occasional backfire.
  - Smells of exhaust fumes and raw fuel.
  - Sight of many types of vehicles, both moving and at rest.
- Lead the animals as close to active firing ranges as safety will allow, near rail yards and crossings, and in areas near operating airfields. These areas get them used to loud, sharp sounds and the sight of flying aircraft. Another way to accustom the animal to gunfire is to crack a bullwhip near it.
- Place cans containing pebbles, boxes of tin cans, or other noisy cargo on the loads so that creaks, rattles, and unusual noises generated on the load will not spook the animal.
- Subject the animals to such odors as iodine, ether, smoke, gasoline, disintegrating flesh, and rotting vegetation.

TRAINING FOR BALANCE

3-81. Units normally use pack animals when crossing over terrain that is impassable by any type of motor vehicle. Traversing such ground requires a well-developed sense of balance. The mule naturally has a fine sense of balance. However, saddle and load will interfere with natural balance unless the animal has some training under load over difficult terrain and learns to adjust to the actions of this dead weight on its back. The handler should train the animal on terrain that is as similar as possible to the terrain in the operational area. He should make the animal cross extremely narrow bridges, fallen trees, and ditches. The handler should work the animal on steep, narrow trails; on corduroy roads; over swamps and boggy areas; and along rocky slopes where it must select safe footing among loose stones. The handler should use judgment in handling pack animals on difficult terrain, and should avoid interfering with the animal's natural balance. The mule usually shows better judgment than the average handler. Therefore, the handler should let the mule pick its way through the very difficult terrain.

Chapter 4

Animal Health Management

Evaluation of the general state of any animal's health is an ongoing process accomplished through daily observation of behavior and routine examination of specific areas of the animal's anatomy. A rudimentary understanding of anatomy and physiology, a few simple examination techniques, and a familiarity with the behavioral patterns of the healthy animal form the foundation of veterinary science. Aided by materials contained in the basic first-aid kit, handlers can treat pack animals suffering from minor injury or disease.

Soldiers performing veterinary care on animals in underdeveloped areas need to keep local cultural practices and economy in mind. An injured or debilitated animal may be of no use to the local people except as a significant source of food. It may be more practical for them to humanely destroy the animal and salvage the meat for food, rather than provide the animal with long-term care. Administration of medications may render the meat inedible. Before initiating any type of medical care involving medications, sedation, or other drugs, the Soldier needs to ascertain whether the animal may be used as a food source; if so, treatment in the form of drugs should be withheld pending the animal's immediate, humane destruction.

ANIMAL BEHAVIOR

4-1. Animals exhibit behavior patterns that resemble human emotion and well-being. They express anger, fear, and boredom with characteristic mannerisms. They also express pain and disability. Regardless of the individual personality traits developed by each animal, general behavioral patterns are common to the species and are useful as tools to evaluate physical status.

HEALTHY ANIMAL BEHAVIOR

4-2. The behavior a well animal exhibits is the benchmark. The handler must become familiar with its behavior patterns and be alert to changes in emotional or physical well-being. The healthy animal looks and acts well, reflecting mannerisms associated with humans. Emotions run the full spectrum from sadness to happiness. Behavior patterns become especially apparent when animals are observed in groups.

4-3. An animal shows its **happiness** by an initial dropping of the head, followed by lifting the muzzle in a circular manner. The upper lip may curl, displaying the teeth in a classic "horse smile." Prancing and lifting the tail usually accompanies this behavior. The well animal, pleased with its activities, may also prance with its ears pointed forward, nostrils flared, tail up, and neck arched with head pointed down.

4-4. Intense absorption, where the nose, eyes, and ears are all intently focused on the item or surroundings, indicates **curiosity**. Even the stance brings the body into line. The animal will sway its head to better see an object directly to the front.

4-5. Circling, pawing or stamping, head shaking, and sideward dancing characterize **frustration**. As frustration continues, the animal's activity results in sweating and unruliness. This behavior also typifies nervousness.

SICK ANIMAL BEHAVIOR

4-6. The sick animal, whether its illness is emotional or physical in nature, exhibits behavior indicative of its condition. An unhappy animal that is otherwise healthy will bring illness on itself. A sick animal will exhibit mannerisms that aid in early awareness of injury and disease.

4-7. Signs of intense interest characterize **fright**. The animal will shift its head from side to side, allowing better vision. It holds its head high and makes audible sniffing sounds. Dancing and shying, often in a circular motion, shivering, and tail swishing accompany increasing nervousness. Its ears will be mobile, seeking the source of threat when it is unapparent. As anxiety increases, profuse sweating occurs. The animal will attempt to flee and, if restrained, may buck and kick. Under extreme conditions, its eyes will roll, showing the whites. Ultimately, the animal becomes terror-stricken. At this point it will scream and, if loose, will attempt to run through any obstruction until it collapses in shock.

4-8. A more common problem is **boredom**. Unchecked, it leads to behavioral problems that will remain a part of its personality. Chewing on wood such as stall doors (cribbing), drawing air into its stomach (wind sucking), and rocking side to side (weaving) are common traits adopted by a bored animal.

4-9. A horse demonstrates **irritation** by laying back its ears and swishing or lashing its tail, often exaggerating the motion on one side. The animal will tense its front or hindquarters and lift a hoof in preparation for kicking. It can glare threateningly and bite or kick with surprising swiftness and ferocity.

4-10. **Pain** elicits reaction according to location (internal or external) and severity. With mild pain (fly bites, chafing, or early arthritis) and hoof pain (such as a pebble caught in the shoe), the animal will shrug, shiver, or kick in an effort to dislodge or soothe the area. It may flick its tail and nip at the afflicted area as a means of removing the irritant. Overall activity is fairly continuous, be it biting and bumping against an object or head tossing in distraction. Head shaking and ear twitching may also represent a head injury or uncomfortable bit.

4-11. An **internal injury** may exist when the animal shows signs of generalized listlessness, melancholia, increased distraction, and flagging (pointing the head) toward the affected side. The animal may lay its ears back and nip or rub its side. It lifts its hind legs alternately and hunches its abdomen in attempts to relieve the discomfort. Vocalizations such as groans are common.

If pain, such as colic pain, continues, the groaning deepens and the animal will bite itself on the side and roll on the ground, worsening the condition.

4-12. An animal shifting its weight to the unaffected side and limping or refusing to move indicates **leg pain**. The animal will also "point" or extend the affected leg to try to relax it. The head will bob or swing to the side in an effort to bring the afflicted side forward.

4-13. Signs of **illness** may include a dulling of the coat, decrease in appetite, or such obvious symptoms as discharges from the eyes or nose, running sores, or diarrhea. The animal will carry its head down and show blankness in its eyes. The legs tend to splay outward while standing and the ears point outward in a splayed fashion also. Progressive illness leads to increased severity in symptoms. Leg splaying, head drooping, eye and coat dullness, fever, shivering, sweating, swaying, and staggering are all manifested to varying degrees.

OVERALL BEHAVIOR PATTERNS

4-14. Through examination of healthy animals and familiarity with their behavior in natural surroundings, handlers gain early recognition of the sick or injured animal. Neglecting small changes in behavior, coupled with ignorance of common maladies, may result in delay of movement or loss of an animal.

PHYSICAL EXAMINATION

4-15. The physical examination of an animal proceeds in the same overall manner as any physical exam. Certain precautions and allowances for the animal's natural fears and curiosity must be exercised. The examiner should enlist help in controlling the animal and constantly remain alert for signs of bolting, biting, and kicking. He should never assume a position between an animal and a fixed object. By leaning or shying sideways or forward, the animal can pin and injure the examiner or helper.

4-16. The animal should be examined when it is quiet and rested. Excitement and heavy exertion will drastically alter the animal's temperature, respiration, and pulse. When the physical examination is combined with routine grooming, the animal will receive it with more cooperation. Also, during every rest period and bivouac, the handler should examine every animal briefly for common maladies.

4-17. The handler should follow an orderly sequence during each examination to preclude overlooking any area. The handler should keep a record of the animal's feeding habits, bowel and bladder habits, general demeanor, and any cough or discharges. He should record results of physical examinations in chart form and include all immunizations. The handler should also maintain an extract version with the animal.

EQUIPMENT

4-18. A general examination may be performed without the use of any medical equipment. However, when available, the handler should use the following items:
- Watch with second hand or second indicator.
- Veterinary rectal thermometer (flexible, digital).

- Petroleum jelly or substitute.
- Stethoscope.
- Penlight or ophthalmoscope.
- Equine dental float (file).
- Rubber gloves.
- Several pieces of cloth or gauze.
- Twitch or hobbles for uncooperative animals.

PROCEDURE

4-19. The examiner begins the examination with an overall look at the animal, noting general demeanor, carriage, and gait. When unfamiliar with the particular animal, he should ask the handler about changes in diet and elimination. Also, he should ask about problems encountered during previous examinations, such as biting or kicking, medical conditions, and injuries. The examiner then checks specific areas of the animal. Many examiners refer to the head, ears, eyes, nose, and throat as HEENT.

Head

4-20. The examiner looks for signs of chafe from the halter, such as lesions and areas of hair loss. He notes how the animal carries its head and if it shakes or rubs it. Encephalitis or concussion and distemper (lockjaw) are detectable by confusion and loss of coordination.

Ears

4-21. The animal's ears should be checked gently because they are sensitive. The examiner looks for halter chafe, lesions (external and internal), and discharge. He notes how the animal carries its ears and watches to see if it shakes its head or rubs its ears. Ear mites, ear flies, and ticks are the main problems encountered.

Eyes

4-22. The examiner performs the eye examination in two parts. First, he does an overall examination of the eye, its orbit, and the lids with available light. He checks for lesions, foreign matter, and discharges. The examiner notes the color of the conjunctiva and the third eyelid—a whitish membrane that closes to cleanse the eye of particles. Communicable eye infections, biting flies, and gnats pose significant problems. Corneal abrasions and ulceration are common, especially in dusty or sandy environments. These minor problems can become serious if not treated properly, leading to blindness or loss of the entire eye. Occasionally, lesions (tumors) will grow on the inner eyelid, requiring surgery. When an animal exhibits photosensitivity by squinting or partially closing an eye, it may be suffering from conjunctivitis or other problems, such as snow blindness. Next, the examiner uses a light source, such as a flashlight, ophthalmoscope, or candle, and working in shade, examines the cornea. By shining the light from an angle, he looks for opacities and surface irregularities. Shining the light from the front, he observes the quality and clarity of the reflected image in the eye. To see if the animal is blind, the examiner moves his open hand in and out toward the animal's eye. If the animal has no reaction to this procedure or does not blink, the animal is most likely blind. The examiner performs the same procedure with the other eye.

Nose

4-23. The nostrils are sensitive and may be moist or dry, according to the environment. Nasal discharge presents a draining moistness, which increases in profusion when the animal's head is down, as in grazing. Discharge is usually accompanied by noise, such as snorting or sneezing, and licking or rubbing to clean the lip and nostrils. Sores are minimal unless inflicted by branches or the lead rope. Animals recovering from pneumonia, strangles, or milder respiratory infections may develop sinus infections. These infections are characterized by drainage, which increases when the head is down. The examiner should tap lightly with a knuckle on the bone just below each eye. Pain will be elicited in an animal with infection.

Oral Cavity

4-24. This examination includes the mouth, lips, gums, teeth, and tongue. The lips may contain splinters or pieces of burrs. The examiner checks for areas of abrasion around the mouth where the coat is rubbed away. Such abrasions may indicate an ill-fitting halter or bit. Malocclusions of the teeth, such as severe overbite or under bite, may interfere with grazing. The examiner checks the teeth for uneven wear, which occurs naturally but is also a common sign of cribbing (gnawing). Normal wear causes uneven, sharp slopes of the molars, whereas cribbing causes uneven wear on the incisors. The examiner checks the inner edges of the lower molars (lingual) and the tongue. When the molars become sharp, the tongue is often ulcerated by contact. He also examines the outer edges (buccal) of the upper teeth. Sharp edges here will ulcerate the inner cheek. In either case, the examiner uses a tooth float (rasp) to smooth the sharp surfaces. Lesions on the tongue can be traumatic and may be caused by ill-fitting bits or plant awns (bristle-like fibers, as on the heads of barley, oats, or wheat). Certain systemic illnesses, such as kidney disease and certain other infectious diseases, can cause ulceration of the tongue and oral mucosa.

Neck and Mane

4-25. The examiner checks the neck and mane for lesions and signs of chafe from contact with tack items. Ticks and mites are frequent in this area because the animal has difficulty dislodging them. Lymph glands located below the jaw can become enlarged from inflammation. Inflammation is somewhat difficult to detect in the glands except in the case of strangles. The examiner takes the pulse using the artery that runs along the inside of the jaw (mandible) on both sides of the head. The pulse will vary from 36–40 bpm during normal rest to 80–100 bpm after exertion, especially at higher elevations. Blood samples can be taken easily from the jugular veins on either side of the neck.

Chest and Shoulders

4-26. The animal's chest and shoulders are examined primarily for lesions caused by foliage and saddle rigging. The examiner notes the carriage, which is the distribution of the weight on the forelegs. He steps back and records the respirations by observing the rise and fall of the chest from the side. The animal must be rested and acclimatized. Normal rates will vary according to species and altitude. The normal rate at sea level is 8 to 16 breaths a minute. After exertion, a rate of 30 to 40 breaths a minute is normal. An animal that coughs or has noisy or staggered breathing requires an examination of the chest (lung fields) with a stethoscope. The examiner listens for gurgling, grating or rustling, or an absence of sound.

Flanks and Abdomen

4-27. The flanks and abdomen receive the heaviest amount of wear from the saddle and the associated rigging and loads. The examiner should thoroughly check for signs of chafe and loss of hair. A lesion in this area will rapidly become worse and possibly become infected without early management. Cysts and boils will require lancing and antibiotic dressings. Relief from further irritation is essential. An animal with an ulcerated area should not be put back under blanket or harness until fully healed and showing hair regrowth. An animal exhibits internal pain, such as colic, the same way a human does, with drawing-up and obvious attention to the afflicted region. The examiner should always know the animal's recent history of appetite and thirst. Loss of appetite is a sure sign of disorder or malady. When colic is suspected, the examiner uses a stethoscope to examine the abdominal region (if colic is present, there will be an absence of normal stomach noises) and supplements with a rectal exam.

Forelegs and Hooves

4-28. The animal's forelegs and hooves can be examined by lifting the lower leg rearward. The examiner supports the leg in one hand and examines the leg from sole upward. He pays particular attention to the entire hoof and each joint of the leg. To check the hoof, the examiner may need a hoof pick or similar device to clean the sole before the examination. All unit members should be familiar with specific injuries common to the legs and hooves. They should also be knowledgeable of specific hoof and bone deformities.

Hind Legs

4-29. The hind legs are best examined from slightly to the side, never directly to the rear. The examiner lifts the leg rearward and checks the hoof. He returns the hoof to the ground and examines the leg from the bottom up. Pulling the tail down and to the side will encourage the animal to stand with both hooves firmly on the ground. Injuries and conformation problems are less common than on the forelegs.

TEMPERATURE

4-30. While holding the tail down and to the side, the examiner takes a rectal temperature, ensuring the thermometer is well inserted and remains in place for at least 1 minute. Alternatively, digital thermometers should remain in place until the thermometer signals that the reading is complete (about 1 minute). Normal temperature for a horse is 99 to 101 degrees Fahrenheit (F). Normal temperature for a mule is 100 to 101.5 degrees F for a mule.

4-31. The proper way to take a rectal temperature is as follows:

- Shake the thermometer down if it is analog (mercury).
- Wipe it clean with alcohol.
- Cover it with petroleum jelly.
- Tie a cord to the thermometer if it has a ring at the end of it.
- Hold on to the cord or tie it to the animal's tail after inserting the thermometer into the animal's rectum.
- Keep the thermometer in for at least 1 minute.
- After withdrawing the thermometer and reading it, clean it with soap and water.

NOTE: If using a digital thermometer, the examiner should follow the manufacturer's directions.

OVERALL OBJECTIVES

4-32. Grooming is an essential part of animal care. The daily grooming period is the ideal time to perform a routine examination, excluding temperature, pulse, and respiration. These are checked only when illness is suspected. Early detection and correction of problems, such as chafe, is essential in the prevention of more serious disorders and possible loss of the animal. The examiner should never overestimate the durability of the animal. A mule is more durable than a horse. However, all animals are vulnerable to a wide variety of problems caused by both the environment and infection passed from animal to animal. Despite great size and strength, these animals are among the most susceptible to injury and disease. Constant vigilance must be exercised to maintain health and effectiveness of animals during sustained operations.

FIRST-AID SUPPLIES

4-33. A basic first-aid set should be carried specifically for use on the animals because standard bandages are too small for most purposes. The handler should put supplies in a weatherproof container in sufficient quantity to care for 10 percent of the animals. As with all critical items, duplication is suggested.

4-34. Figure 4-1, pages 4-8 and 4-9, lists the minimum supplies and equipment needed for treatment of a wounded animal. They include dressing changes since bandaging supplies (tape, cotton, gauze) will be expended rapidly and will require restocking or supplementing.

4-35. The examiner should include and apply heavyweight sutures with the same considerations as in humans. More sophisticated procedures require additional supplies. In most instances, the same equipment carried for treatment of humans is applicable to treatment of animals. Antibiotics and steroidal anti-inflammatory medicines are a notable exception. With these agents, it is either sensitivity to the drug or altered prescribing principles that pose the exception. A veterinarian must approve all medications before including them in the kit. Because replenishment of veterinary supplies may be difficult, the handler should include possible substitute or expedient materials, resupplies, and caches in mission planning.

Item	Quantity
Providine iodine solution	500 milliliter
Chlorhexidine solution	500 milliliter
Providine iodine or chlorhexidine surgical scrub brushes	2 each
Sterile saline for irrigation (0.9%), 500 milliliter	1 each
Lidocaine 2%, injectable 100 milliliter	1 each
Triple antibiotic ointment, 15-milliliter tube	1 each
Xylazine, injectable, 100 milligram/milliliter	100 milliliter
Trimethoprim sulfa, 960-milligram tablets	200 tablets
Duct tape, 100-mph tape	1 roll
Gauze, 4- x 4-inch, nonsterile	1 package
Vetwrap/Coban, 4-inch roll	6 rolls
Cotton, 12-inch roll	1 roll
Elastic bandage (Elastikon), 4-inch roll	2 each
Brown gauze, 6-inch roll	4 each
Conforming gauze bandage, 6-inch roll (for example, Kerlex)	3 rolls per animal
White bandage tape, 2-inch roll	2 rolls
Gloves, latex, exam, nonsterile	100 pairs
Gloves, latex, surgical, size 7 1/2	4 pairs
Stethoscope	1 each
Penlight	1 each
Scalpel blades, #10	6 each
Scalpel handle, #10	1 each
(Alternate to above 2 items: Disposable #10 scalpels with handle)	10 each
Minor Surgery Kit	
Scissors, surgical (Mayo or Metzenbaum)	1 each
Forceps: Brown Adson or Allis tissue forceps	1 each
Large needle holders with scissor blade	1 each
Hemostats, curved	4 each

Figure 4-1. Supplies and Equipment Needed to Treat a Wounded Animal

Item	Quantity
Surgical towels	3 each
Gauze pads, 4- x 4-inch, sterile	10 pads
Gauze pads, 24 to 30 inches, sterile	3 sheets
Vetwrap/Coban, 4-inch roll	1 roll
(Package this kit together and autoclave.)	
Leg Bandage Kit	
Sheet cotton, 26- x 30-inch	3 sheets
Vetwrap/Coban, 4-inch roll	3 rolls
Brown gauze, 6-inch roll	1 roll
(Package this kit together in a plastic bag and compress with food vacuum preserver. The kit will compress down to a very compact size. This kit has essential items for one lower limb bandage. For full limb, two kits are used.)	

Figure 4-1. Supplies and Equipment Needed to Treat a Wounded Animal (Continued)

FIRST-AID TREATMENT

4-36. Most principles of care in the administration of first aid directly parallel the treatment of humans; the major differences are in anatomical structure. Ethics will play a lesser role in treatment in the field: It is more likely animals will be destroyed in the field even though they could have been saved under normal circumstances by performing a radical procedure. Otherwise, the time involved for an animal to recover from lameness or the debilitation that occurs as a result of a serious injury could hamper the unit's ability to accomplish its mission.

OPEN WOUNDS

4-37. Bleeding, infection, and tissue loss are the main concerns in wounds resulting from external trauma. The examiner should give particular attention to any open wound on the hoof or hind leg. These wounds pose the highest threat of contamination and subsequent infection from bacteria contained in feces. Horses and mules are very susceptible to tetanus. The examiner should supplement treatment of open wounds, especially punctures, with a tetanus booster. Wounds directly over or close to joints may actually penetrate into the joint. Severe lameness or arthritis can result from joint infection, and these wounds need to be treated aggressively with systemic antibiotics.

LACERATIONS

4-38. These wounds, caused by a tearing of the skin, usually result in slight to moderate bleeding. The vessels contract and limit the flow of blood to the affected area. The examiner inspects the injury to determine the severity, depth of penetration, and degree of damage to underlying structures. He cleans it with soap and water to remove all foreign material. When the wound is small and the edges will remain closed, he applies an antibiotic ointment and covers the wound with a clean dressing. The examiner closes the larger wounds, and those where the edges gape open, with sutures, butterfly bandages, or surgical staples, if available. If closing the skin will leave a large pocket or cavity under the skin where tissues are damaged, it is better to leave the wound open than to suture the skin. Large cavities under the skin tend to accumulate blood and body fluids and can promote infection. The examiner applies antibiotic ointment and a clean dressing. He keeps the area dry and changes the dressing daily.

INCISIONS

4-39. Incised wounds tend to bleed freely. The first priority is to **STOP THE BLEEDING**. The examiner should use direct pressure, pressure dressings, tourniquets, and ligation with the same indications and precautions as on humans. Circulation to the lower leg is mainly superficial. Direct pressure is usually sufficient to stop bleeding in the extremities. Consequently, the examiner uses a tourniquet only when absolutely necessary and releases it every 15 minutes. Ligation should be accomplished as soon as practical to restore circulation to the limb. When the wound is dirty, the examiner places a constricting band above it to reduce bleeding and quickly, but thoroughly, cleans the wound before closing it with a dressing. The examiner uses sutures to close large or gaping incisions. When available, he gives a tetanus booster.

PUNCTURE WOUNDS

4-40. The causative agent may remain in the wound or be withdrawn. In all cases, the first step is to **STOP THE BLEEDING**. The examiner evaluates the severity of the wound and continues to control the bleeding. Puncture wounds are either low or high velocity. Low-velocity wounds impact only on the actual tissue penetrated and tend to bleed freely. High-velocity wounds impact on the actual tissue penetrated and impart a shock wave into surrounding tissue. This concussion tends to cause contraction of ruptured vessels, temporarily reducing the immediate blood loss. Packing the wound with gauze will aid in stopping blood loss when done in conjunction with standard pressure dressings. Impaled objects can also aid in stopping blood loss and may be left in place for a short period after being trimmed off flush and supported with a dressing. However, if the object impedes joint function, it must be removed. A topical antibiotic should never be put into a puncture wound; such an action causes systemic absorption of a topical medicine. The examiner irrigates the wound thoroughly with saline or water, sterile if possible. He cleans and dresses the wound. He then administers a tetanus booster, if available, especially in cases where punctures could allow direct access of pathogens into the bloodstream.

4-41. Animal handlers should follow specific steps when dressing wounds. The procedure for applying routine dressings is as follows:

- Inspect the wound and take appropriate measures to stop the bleeding.
- Cleanse the wound and surrounding area with clean, warm water and a nonirritating antibacterial soap. Typically, the area surrounding the wound should be shaved during cleansing.
- Apply a triple antibiotic ointment or other antibiotic topically.

- Cover the wound with gauze.
- Cover the gauze dressing with cotton padding.
- Wrap the area with roll gauze and secure with tape.

4-42. Animals will tear off the dressing if it is loosely applied. There are roll tapes specifically intended for use with animals. This tape (Vetwrap, similar to Coban bandage for humans) sticks only to itself, diminishing the trauma of dressing changes. Also, the handler should be careful of overtightening the dressing, creating a tourniquet.

CLOSED WOUNDS

4-43. Closed wounds result from external mechanisms such as overuse, hyperextension (straightening past the normal locking point of the joint), or hyperflexion (bending beyond the normal range of motion). Injuries include bruising, stretching, or tearing of connective tissue; joint dislocation; bursal inflammation or rupture; cartilage damage; and various degrees of bone fracture. Symptoms of these injuries are swelling, stiffness, and a partial or complete loss of function.

4-44. Major partial- and full-thickness burns and displaced and open fractures require specialized, lengthy treatment and recuperation. Destroying the animal becomes a matter of operational necessity when conducting a mission.

ALLERGIC REACTIONS

4-45. Allergy or anaphylaxis (severe allergic reaction) stems from a variety of causes. Most common are bites, stings, and skin absorption. Less common, but more serious, are reactions caused by substances eaten by the animal and entering the bloodstream. Horses, donkeys, and mules can also have allergic reactions to certain medications and vaccinations. Signs range from mild urticaria (hives) to anaphylaxis.

Bites and Stings

4-46. Allergic reactions to insect bites and stings result in small blistered areas (wheals) or generalized swelling (edema). The extent of swelling is dependent on the number of bites or the toxicity of the venom injected. A single fly bite will result in a wheal, whereas multiple bee stings will result in generalized swelling. Snakebites and scorpion stings produce similar reactions. Treatment is symptomatic. The handler applies ice to the affected area to reduce swelling and slow the spread of venom. Generally, large animals tolerate venom better than humans. However, antihistamines increase the effect of snake venom and **SHOULD NOT** be administered. The handler gives tetanus antitoxin and broad-spectrum antibiotics to counteract bacterial infection associated with snakebites. Emergency tracheotomy has been successfully used in cases of respiratory distress. Constricting bands will slow dissemination while increasing local necrosis and circulatory obstruction. Treatment for shock is indicated in moderate to severe cases.

Urticaria (Hives)

4-47. Stinging nettle, poison ivy, and chemical irritants result in blistering or rash formation. These rashes progress rapidly from localized to generalized when an animal becomes sensitized from repeated or heavy exposure to the irritant. Cold soaks and antihistamines will reduce the reaction, especially when respiratory distress develops. Ingestion of a large amount of protein will produce a similar effect. Gorging on food concentrates, spoiled hay, and a variety of other ingested substances will result in an allergic reaction (anaphylaxis). To treat the symptoms, the handler passes a tube to remove the stomach contents and gives laxatives and antihistamines.

BURNS

4-48. Burns are extremely painful injuries for animals and are difficult to manage if severe. With full-thickness burns over more than 30 percent of the animal's body, or over the head and face, prognosis is poor and euthanasia should be considered.

Superficial Burns (First-Degree Burns)

4-49. These burns are characterized by reddening of the skin. Vesicle (blister) formation is uncommon. Localized tenderness will necessitate refitting of tack or shifting loads until the animal tolerates pressure on the afflicted area. No other treatment is necessary.

Partial-Thickness Burns (Second-Degree Burns)

4-50. Tissue damage from these burns is limited to superficial layers of the skin. Vesicle formation occurs, resulting in the peeling away of dead skin after the fluid has drained and the blister dried. Pain lasting several hours after initial injury subsides to localized tenderness. Treatment of the burn consists of trimming the coat over the afflicted area and examining the wound. The handler applies an appropriate topical medication (such as silvidene ointment). He covers the area for 3 days with a clean, dry dressing to prevent vesicle rupture. Additional padding is required under harness straps when the animal must be harnessed. The handler drains all vesicles on the fourth day and allows drying for 5 days. He continues to protect the wound during this time to prevent infection. He trims away the dried skin on the tenth day after initial injury. The handler then places padding under tack until healing is completed.

Full-Thickness Burns (Third-Degree Burns)

4-51. These burns are characterized by charring of the flesh and coat. Severe tissue damage extending into underlying layers occurs. Moderate to severe pain, dehydration, and shock are symptoms. Treatment, when small areas are involved, is generally successful. However, severe burns covering a large percentage of the animal require clinical attention or destruction. The handler treats in the same way as for second-degree burns, encouraging fluid intake, adequate rest, and a restricted diet. Topical burn preparations will lessen pain; however, novocaine (lidocaine) nerve blocks may be required for the first days. The handler observes for signs of dehydration, infection, and shock.

LAMENESS

4-52. This general term describes many types of injury and conformation faults. Basically, lameness is caused by pain in one or more legs. Conformation faults, arthritis, and a variety of joint and tendon malfunctions result in lameness. Often in these cases, first-aid treatment is not applicable. Consequently, only lameness resulting from injury will be covered in this section.

4-53. The handler examines the animal in motion to isolate the limbs involved. The animal will step quickly off a painful leg and swing toward its good side. The handler observes for stiffened joints, reduced range of motion, nodding, limping, or swaying gait. He examines suspect limbs from the sole upwards and always looks for a hoof injury first. He does the examination in three parts—inspecting the sole, tapping the hoof, and flexing the individual joints, watching for evidence of pain. The handler walks the animal after each examination. If the gait is affected, he focuses on the last area examined. This procedure is time-consuming but is the only practical method for field use.

Hoof Lameness

4-54. The hoof is vulnerable to a variety of injuries and infection. An unshod, untrimmed hoof will develop cracks. Injuries from overtrimming and improper nailing occur during routine trimming and shoeing. An animal can develop corns if it goes too long without reshoeing. While moving, impact forces cause stress on the bones, suspensory ligaments, and tendons. Rocks and a multitude of sharp objects bruise and pierce the sole or lodge under the shoe. These are injuries for the farrier's or the veterinarian's care. Exceptions are thrush and cracked heels as explained below.

4-55. Routine care of the hoof consists of regular trimming and shoeing. During halts, the handler inspects the sole and removes any objects that have lodged there, especially in rocky terrain. Painting the wall (outside) of the hoof with hoof oil may reduce cracking.

4-56. **Thrush** is caused by bacteria infecting the frog. It is characterized by a dark, foul-smelling discharge from between the frog and the sole. Most common among stabled animals, it is caused by standing in wet and unsanitary conditions, or from untrimmed hooves where the frog has become overgrown. Treatment consists of proper hoof trimming, thorough cleansing supplemented by application of iodine, copper sulfate, or a diluted bleach solution to the infected area around the frog only.

4-57. **Cracked heel** is actually a skin condition affecting the pastern above the heel of the hoof. Caused by continued incomplete drying of this area, it is seen in animals pastured in wet or muddy fields. The skin develops dry scaling that degenerates into cracking, as in athlete's foot. Treatment consists of cleansing and thorough drying. The handler applies an ointment or petroleum jelly. The problem is a result of dryness and chafing. The animal should never be treated with an agent that will cause further drying of the area.

Lower Leg Lameness

4-58. Any injury below the knee is potentially very serious. There is little muscle tissue in this area, the blood supply is limited, and the area is in continual motion, which tends to aggravate injuries. Furthermore, the chance of infection is greater in areas close to the ground because they are hard to keep clean.

4-59. Injury to the lower leg is a result of either stress, caused by a misstep or prolonged travel on a hard surface, or trauma from hoof strikes inflicted by other animals or self-inflicted. Isolated occurrences result in temporary lameness. Repeated or prolonged injury generally results in formation of calluses or calcium deposits in the area. Treatment of isolated occurrences is successful when time for adequate rest is allowed. Formation of calluses and calcium deposits require care by a veterinarian and protracted periods of rest.

4-60. When, as the result of an inflammation, synovial fluid collects around a joint or tendon, the condition is termed synovitis. Synovial fluid is produced as a lubricant for joint motion. Inflammation will cause an overproduction of this fluid, which is then trapped in and around the joint. Reabsorption occurs naturally over time in most cases, provided inflammation is reduced. Needle aspiration and cortisone injections are stopgap measures and may result in the eventual destruction of the joint. With early diagnosis and prompt treatment, these conditions need not result in lameness. Treatment consists of rest, ice packs, or cold soaks and reduction of load or variations in surface traveled. Some of these cases will result in permanent deformity in the form of an enlarged joint capsule.

4-61. Sprains of the flexor tendons or suspensory ligament occur during running or jumping but may be seen when tripping occurs on a wedged hoof. A sprain is a stretching, tearing, or complete rupture of the affected tendon or ligament. A common name for these conditions is "bowed tendon." Symptoms consist of lameness, which is often severe, and pain and swelling over the injured region. Inflammation will generate warmth, detectable to the touch. In severe cases (rupture of the tendon), loss of support occurs in the joint during weight bearing. Length of recovery depends on the extent of the stretch or tear.

4-62. Treatment consists of complete rest, ice packs (at least twice daily), or cold soaks, supplemented by supportive wrapping or casting of the leg. Recovery from a moderate tear or complete rupture requires surgical repair. Needle aspiration and cortisone injections have promoted healing in some cases.

PARASITIC INFESTATION

4-63. Parasites are classified according to location of residence: external (ectoparasites) or internal (endoparasites). Of the two, internal parasites are more debilitating, though it is external parasites that carry the most infectious disease and present the greatest nuisance. There are large numbers of different parasites. Of these, some affect only a specific host. Others afflict any warm-blooded animal, including man.

ECTOPARASITES

4-64. External parasites that reside on or just under the surface of the skin by burrowing include insects such as flies, fleas, lice, mosquitoes, mites, and ticks. The area around the eyes, ears, neck, and anus are the most common sites of infestation because they are areas of secretion, and the skin in these regions offers the easiest penetration.

4-65. Depending on the type of parasite, symptoms range from rashes and blistered areas to patches of hair loss (alopecia). Animals displaying excessive itching, hair loss, or rough, thickened skin should be suspected of hosting parasites. Closer examination will reveal blisters or burrows, dried blood, or the insect itself.

4-66. The handler treats for parasites by applying topical insecticides. This treatment works best on those parasites that remain on the host, such as mites. It is less effective on flying insects because the brief feeding period reduces total dosage absorbed. Insect repellents provide relief from these pests and are the only practical solution in a field environment.

ENDOPARASITES

4-67. Internal parasites generally live within the intestinal tract. They remain there throughout their life cycle or migrate out the anus as larva to spend their adult cycle outside the host. Less common are the endoparasites that live in other internal organs or surrounding muscle tissue. Regardless of residence, the mouth is the most frequent point of entry.

4-68. Symptoms are subtle, except for the dramatic appearance of the parasite exiting the anus or contained in manure deposited on the ground. Evidence of unexplained weight loss, chronic tiredness, or dullness of coat is frequently the result of parasites. These anemic-like symptoms result from the loss of nutrients to the parasite, especially a large colony of parasites. Untreated, these colonies debilitate the host, often causing blockages and colic.

4-69. Treatment of internal parasites is more successful than treatment of external parasites because a more consistent dosage of antiparasitic is absorbed by the invading parasites. Preventive administration of antiparasitics is a common practice and is the most convenient method. Many anthelmintic medications are available for horses, donkeys, and mules, including ivermectin, fenbendazole, oxibendazole, and moxidectin. They are available in an easy-to-administer oral paste form. When treating animals that have not been dewormed regularly throughout their life, the first treatment should be given as half the normal dose, 2 weeks apart. With extremely heavy parasite load, rapid die-off of large numbers of parasites can cause intestinal impaction, diarrhea, or other adverse reactions.

DISEASES

4-70. Zoonosis are diseases that usually affect only animals but can be transferred from animals to humans under natural conditions. Anthrax (cattle fever) and rabies are excellent examples. People working with animals are at particularly high risk for zoonotic diseases.

4-71. Currently, there are immunizations for most major infectious diseases affecting horses, mules, and similar animals. The key to disease management is to follow a schedule of routine inoculation and parasite control. Additionally, proper sanitation and hygiene must be maintained.

4-72. New and ill animals should be quarantined to prevent the spread of disease among the healthy animals. Frequently, animals will develop flu-like symptoms after being transported or stressed by changes in environment.

This syndrome is referred to as "shipping fever." Disease pathogens congregate in food, bedding, and tack. Therefore, personnel should ensure that these items remain clean and are not transferred from animal to animal without some form of disinfection.

STRANGLES

4-73. Strangles is an infection of the lymph nodes caused by the bacteria *Streptococcus equid*. This condition is highly infectious between horses and is usually seen as an outbreak within a herd or stable.

4-74. Occurrence is most frequent among young animals but will occur in any equine not previously exposed or immunized. The infectious period is approximately 4 weeks. The incubation period is 3 to 6 days, with flu-like symptoms. Fever with temperatures as high as 106 degrees F may occur. Few infections in horses cause as significant fevers as with *Streptococcus equid* infection. If a fever over 103.5 degrees F is observed, strangles or equine influenza should be suspected. A nasal discharge follows rapidly and is quite heavy. As inflammation of the lymph nodes in the throat and neck continues, and abscesses form. Abscesses can also form in any lymph nodes of the body, especially in the mesenteric lymph nodes along the intestines. The animal may also have a foul smell about it. These are classic symptoms of the disease. The lymph tracts, which run the length of the neck and trunk bilaterally, become so swollen that they appear rope-like under the skin. Central nervous system (CNS) damage occurs in some cases. However, pneumonia is the greatest cause of death.

4-75. Treatment is largely symptomatic. The focus of treatment should be on draining abscesses and controlling fever. Although *Streptococcus equid* responds well to penicillin antibiotics, use of antibiotics to treat the disease may result in prolonging the problem. Antibiotics are indicated if secondary pneumonia develops during the course of the disease. Otherwise, with good drainage and nursing care, horses will often completely recover within 4 to 6 weeks.

4-76. Using hot packs to bring the abscess to maturity is more appropriate. Next, the handler should incise and drain (I&D) these mature abscesses using proper techniques to prevent the invasion of other pathogens into the lymphatics. Vaccinations are available to help prevent the disease. Due to the short incubation period (3 to 6 days), once one animal in a herd or stable is affected, it may be too late for vaccination of the other exposed horses to be effective.

4-77. If adequate drainage of abscesses is obtained and fever is controlled, animals can recover completely. However, complications can occur in severe or untreated cases including laminitis, severe myopathy (muscle damage), pneumonia, or airway obstruction. Because of the location of affected lymph nodes in the throat area, the airway can become obstructed in severe cases. Emergency tracheotomy may be necessary in these cases.

TETANUS

4-78. An acute infectious disease, tetanus is caused by an introduction of contaminated soil into tissue and the bloodstream and affects the CNS. This anaerobic neurotoxin is normally inactive and lives in a spore state. Usually introduced through a wound, the disease causes tissue decay, which provides the anaerobic environment required for reproduction. After reproduction, the bacteria rupture, resulting in the release of the neurotoxin, which migrates along the nerves to the spinal cord.

4-79. Symptoms of tetanus, either ascending (motor nerve to spinal cord) or descending (lymphatic to CNS), consist of a characteristic muscle spasm after even mild stimulation and localized stiffness increases to generalized rigidity. Especially prone to spasms are the muscles in the jaw and neck, giving rise to the common name of lockjaw. Symptomatic progression leads to rigidity of the ears, spine, and legs. The stance widens (sawhorse) and the nostrils dilate. Closing (prolapse) of the third eyelid, profuse sweating, continued excitation spasms, rapid respiration, cardiac irritability, and arrhythmia are present in the latter stages.

4-80. Treatment is begun by preventive immunization of all animals, followed by routine booster injection (tetanus toxoid). After an animal is wounded, it should receive good wound care, consisting of thorough cleansing and disinfecting of the wound site, followed by a tetanus booster. In animals that have contracted the disease, treatment consists of drainage and disinfection of the wound, removal of all dead tissue, and an injection of tetanus toxoid and tetanus antitoxin (100 to 200 units per kilogram [kg] intramuscular [IM]) in different injection sites. Treatment with sedatives (acepromazine 0.05 milligrams [mg]/kg IM every [q] 6 hours [hr]) or phenobarbital (6 to 12 mg/kg intravenous [IV], initially followed by the same dose orally q12 hr) has succeeded in moderately severe cases. Procaine penicillin G should be administered (22,000 to 44,000 mg/kg IM q12 hr for 7 days). The handler supplements treatment by keeping the animal in a darkened, quiet stall. He should avoid any incidence of startling the animal. The handler elevates food and water since the animal has difficulty lowering its head. Recovery periods average 2 to 6 weeks.

EQUINE INFECTIOUS ANEMIA

4-81. Commonly known as swamp fever, equine infectious anemia is a viral disease very common among horses worldwide. Transmission occurs from blood-to-blood interaction such as in the use of contaminated syringes or scalpels. It can reach epidemic proportions when transmitted by blood-sucking flies.

4-82. The disease is characterized by flu-like, low-grade fevers, yellowing of the gums, depression of appetite and demeanor, weight loss, and obvious signs of anemia during microscopic blood examination. Continued weight loss, enlarged spleen, swelling of the infected area, debilitation, and death follow if the disease is untreated. Diagnosis through the use of a serological test (Coggins test) is done in the clinical environment. This test shows the presence of antibodies in the blood of an infected animal. Vaccines exist although their effectiveness is questionable. Quarantine of any suspect or new animal and symptomatic treatment is the only therapy. Control of vectors, by use of insect repellents and insecticides, plus proper sterilization of medical instruments will minimize the impact of this disease on the herd.

COLIC

4-83. Colic is a term used to denote pain in the abdomen from a variety of different causes. Colic is often caused by distention of the bowel resulting from excessive gas production (flatulent colic), impaction of feces or bowel obstruction from colonies of intestinal parasites (obstructive colic), twisted intestine (torsional colic), or gorging or overfeeding. Colic may also result from circulatory problems due to the inactivity of bowel segments. Colic from intestinal impaction often occurs when horses do not drink sufficient water. This often occurs at the change of seasons when the temperature suddenly becomes cold or hot, or when horses are moved to an area with a different source of water unfamiliar to them.

4-84. Regardless of the type of colic and its cause, the animal exhibits a sudden loss of appetite, depression, and frank attention to the abdominal region. Bowel sounds frequently diminish or alter. Rectal examination may locate the obstructed region. Marked distention of the flanks may be present, especially in young animals or in severe stages.

4-85. Treatment is largely symptomatic. It includes administering analgesic agents and making sure the animal is sufficiently hydrated. The handler should keep the animal on its feet to reduce the chance of complications, but if the animal is violently painful to the point of hurting itself or humans, it is best to try to allow it to lay down in a soft, open area, such as on sand or grass, with no fences or other objects nearby. A painful horse can throw itself down quite suddenly, resulting in severe injury of its handlers. All nonessential personnel should stay away from the animal. The handler passes a bloat tube (nasogastric) to relieve gastric distention. If the animal does not reflux a net amount of gastric contents through the nasogastric tube (more flowing out than amount of water pumped in) and has no obvious signs of complete obstruction, mineral oil (2 to 4 liters) can be given by stomach tube to disperse impactions. The handler may supplement the mineral oil by administering an osmotic laxative (magnesium sulfate) or an irritant (neostigmine). Passage of the tube and administration of **any** material through the tube should only be performed by personnel well trained in this procedure. Accidental passage of the tube into the trachea instead of the esophagus can result in death of the animal. Analgesics, to prevent self-injury, may be indicated. Flunixin meglumine (Banamine) is recommended at 1.1 mg/kg IV or IM q12 hr. Flunixin meglumine is very effective at relieving mild to moderate pain and reducing fever and inflammation associated with irritation in the intestine. If flunixin is not available, phenylbutazone can be used instead, also at 1.1 mg/kg q12 hr. Butorphanol can be used for severe pain at 0.01mg/kg q1–2 hr IV as needed to control pain. Due to possible excitatory reactions, morphine is not generally used in horses, mules, or donkeys. Decompression through the use of a large-bore needle (trocharization) inserted into the upper flank is sometimes effective in relief of severe distention as a last-ditch effort. Severe cases of colic, such as with intestinal torsion, require surgery to correct the problem. If complete obstruction occurs, pressure from buildup of gas, fluid, and fecal material will eventually cause rupture of the stomach or intestine. If this occurs, the animal will often suddenly become significantly more comfortable for a short time. However, within 30 minutes to an hour, signs of shock, such as trembling, diffuse sweating, staggering, or collapse, will occur and the animal will die soon after. This condition is essentially untreatable and the animal should be humanely destroyed. Prevention, through proper diet, dental care, and parasite control, is the most effective method of dealing with colic.

HYPOTHERMIA

4-86. Hypothermia will kill livestock as well as humans. Shivering is one of the first signs of cold stress. The symptoms progress through listlessness and loss of feeling in the limbs until death occurs.

4-87. Hypothermia is a lowering of the body's temperature. At a rectal temperature of less than 28 degrees Celsius (C) (82 degrees F), the ability to regain normal temperature naturally is lost, but the animal will continue to survive if external heat is applied and the temperature returns to normal. It is important to observe and measure the vital signs—pulse, breathing, mental status, and rectal temperature.

4-88. Knowing the severity of hypothermia is valuable to decide the rewarming techniques to be used for treatment. On the basis of body temperature, hypothermia can be classified as mild—30 to 32 degrees C or 86 to 89 degrees F, moderate—22 to 25 degrees C or 71 to 77 degrees F, and severe—0 to 8 degrees C or 32 to 46.5 degrees F. There is tremendous variability in physiological responses at specific temperatures among individuals and species (Table 4-1). The simplest way to determine whether the patient is hypothermic or not is to assess body temperature by placing a bare hand against the skin (preferably in axilla or groin region) of the patient. If the skin feels warm, hypothermia is unlikely. Patients with cold skin should have rectal temperatures taken with a low-reading thermometer.

Table 4-1. Temperature Ranges for Work Species

Species	Low Dangerous Range (C)	Normal Range (C)	High Dangerous Range (C)	Quit Work (C)
Horse	30–32	37.5–38.5	39.5–40.0	40.5
Donkey	30–32	37.6–38.0	39.5–40.0	40.5
Mule	30–32	37.5–38.0	39.5–40.0	40.5
Ox	31–33	38.0–39.0	39.5–39.8	40.0
Camel	29–33	36.5–42.0	42.5–43.5	44.0
Llama	30–32	37.2–38.9	39.2–40.0	41.0
Elephant	Not Available	36.0–37.0	38.0–38.2	38.3
Reindeer	Not Available	38.0–39.0	39.2–39.4	39.5
Dog	31–33	38.9–39.9	40.9–41.9	42.3
Goat	Not Available	38.6–39.6	40.6–41.6	42.0

EXAMINATION FOR HYPOTHERMIA

4-89. To examine a hypothermic animal, the handler should make sure that the animal has an open airway, is breathing, is within the normal pulse and respiration range (Table 4-2), and has a normal rectal temperature. These factors are commonly known as the ABCs. The handler should also perform a—

- Brief history (for example, duration of exposure regarding circumstances in which animal is found).
- Brief physical examination, to include the following:
    - Feel-of-body temperature.
    - Level of consciousness and neurological examination.
    - Cardiopulmonary examination.
    - Associated trauma.
    - Weight of animal. Depending upon the availability of staff and equipment, chest X ray, urinalysis, complete blood work, and arterial blood gases are also recommended.

Table 4-2. Pulse and Respiration Ranges for Work Species

Species	Normal Pulse (per minute)	Normal Respiration (per minute)
Horse	23–70	12–14
Donkey	40–56	14–16
Mule	35–67	13–15
Ox	60–70	30–32
Camel	40–51	10–13
Llama	60–90	10–30
Elephant	24–40	6–15
Reindeer (Summer)	20–40	20–50
Reindeer (Winter)	40–70	8–16
Dog	70–120	18–34
Goat	70–80	16–34

MANAGEMENT OF HYPOTHERMIC ANIMALS

4-90. Once it is established that an animal is hypothermic, the primary goal in the treatment and handling of the animal is to keep the animal alive by warming, avoid any further exposure to cold, and then transport the animal to a site of complete veterinary care if possible. The severity of hypothermia will determine the rewarming technique to be used for treatment.

HYPOTHERMIC REWARMING TECHNIQUES

4-91. There are three rewarming techniques—passive external, active external, and active internal—which should be used according to severity of hypothermia. To treat the hypothermic animal appropriately, the handler should first know that the animal is in fact hypothermic. Once the severity is determined, the handler has to decide the rewarming technique to be used for treatment.

Passive External

4-92. The animal's own metabolic process continues to produce heat spontaneously, so no external heat is required. Shivering is an example of thermogenesis. This method is the simplest and slowest but is sufficient for mild hypothermic animals.

Active External

4-93. This method includes warm baths, hot water bottles, blankets, heating pads, and radiant heaters. This type of rewarming is safe only for mild hypothermia because externally applied heat stimulates peripheral circulation.

NOTE: The handler should avoid direct application of hot objects or excessive pressure (for example, uninsulated hot water bottles or tourniquets). He should ensure that items such as oxygen and fluids coming into contact with the animals are warmed. A severely hypothermic animal should not be put in a shower or bath.

Active Internal

4-94. These rewarming methods are usually more complex and need to be carried out by professionals (veterinarians or animal health technicians). These include inhalation rewarming (ventilation of patient with heated, humidified air or oxygen), circulation of heated fluids (40.5–43.5 C) in body cavities (gastric, thoracic, and peritoneal lavage), and heated intravenous solutions, preferably dextrose, as these provide energy to meet increased metabolic demands but contribute little heat due to vasoconstriction in cold extremities.

NOTE: Inhalation rewarming is the only method that can be used by a layman and does not require much training (mouth-to-mouth breathing). Inhalation of warm, saturated air delivers heat directly to the lungs and heart. The brain is also warmed from this blood flow and from conductive heat flow from the respiratory and nasal cavities. This method also assists in rehydration as an added benefit.

HEAT AND SUN STRESS

4-95. The range of temperature in which heat distress will be noted is similar to that for humans. If the handlers and other detachment members are showing signs of severe discomfort, the animals may be in distress as well. Direct sun all day long not only overheats animals in most environments, but it can also lead to sunburn. In severe cases, heat exhaustion or sunstroke can result. The handler can use any piece of shade that is available. Livestock should always be placed where a breeze will help with the cooling process. Portable shade can be obtained by putting straw hats on animals' heads.

4-96. The danger signs of heat distress, in all species except camels, are tremors, "drawn" expression, heavy perspiration, panting, frothing at the mouth, and dark yellow urine with a strong smell. Heat injuries begin with "thumps," progress to heat cramps, and advance to heat exhaustion.

- *Thumps* is characterized by panting, widely dilated nostrils, heavy sweating, and a deep thumping sound from the lower lungs and flanks.
- *Heat cramps* occur due to loss of salts, especially potassium. The animal then becomes stiff in the large muscles, such as the hind legs. The animal may cry in pain and try to either stand still or roll on the ground.
- *Heat exhaustion* is shown by large amounts of sweat, tremors of the limbs, and profuse amounts of sweat. If the animal stops sweating, sunstroke is underway.

4-97. In all heat injuries, the animal should be cooled as quickly as possible by pouring cool water over the head and back. Alcohol will cool even faster and should be applied if available. All heat injuries require prompt attention, but sunstroke is a true emergency and requires immediate action or the animal will die.

4-98. Rehydration should be attempted if the animal can drink. If available, citrus-based electrolyte solutions such as a mixture of water, a small amount of sodium salt, and an equal amount of potassium salt flavored with sugar will suffice. This amount is enough to give a slight taste of salt and a stronger taste of sugar. A quick sample of the solution will tell the handler if it is correct. Honey and water with a little salt can also be used. The handler can also add potassium salt if possible.

4-99. Once the animal's body temperature has begun to drop (according to the rectal thermometer), chills must be avoided. The animal should be given a gentle massage to help move blood to the skin. A dry blanket should be applied and secured with a strap. Once this crisis has passed, rest is best under the watchful eye of a veterinary nurse or first-aid provider. If chills set in, the handler should add more blankets and continue to massage the animal. A hot or recovering animal should not be fed grain. It can be given forage and free access to water.

IMMUNIZATION SCHEDULE

4-100. The following paragraphs list several diseases that plague animals, and also provide a suggested immunization schedule. For primary immunization, an initial vaccination is required, followed by a repeat dose in 3 to 4 weeks. Table 4-3, pages 4-22 and 4-23, is a handy reference guide for scheduling immunizations.

Table 4-3. Immunization Schedule

Type	Frequency Profile
Tetanus	All horses; foals at 2–4 months; annually thereafter. Brood mares at 4–6 weeks before foaling.
Encephalomyelitis (WEE, EEE)	All horses; foals at 2–4 months; annually in spring thereafter. Brood mares at 4–6 weeks before foaling. *Note:* Vaccinate for Venezuelan Equine Encephalomyelitis (VEE) in endemic areas only.
Influenza	Most horses; foals at 3–6 months, then every 6 months. Traveling horses that will contact other "outside" horses, every 6 months. Brood mares biannually, plus booster 4–6 weeks prefoaling.

Table 4-3. Immunization Schedule (Continued)

Type	Frequency Profile
Rhinopneumonitis	Foals at 2–4 months and younger horses in training. Repeat at 2- to 3-month intervals. All brood mares at least during 5th, 7th, and 9th months of gestation.
Rabies	Foals at 2–4 months; annually thereafter.
Strangles	Foals at 8–12 weeks; biannually for high-risk horses. Brood mares biannually with one dose 4–6 weeks prefoaling. Some adverse reactions are associated with this vaccination. *Note: Vaccinate only in situations where infection is likely.*
Potomac Horse Fever	Foals at 2–4 months; biannually for older horses. Brood mares biannually with one dose at 4-6 weeks prefoaling. *Note: Vaccinate in endemic areas only.*
West Nile Virus	Initial vaccination, booster in 2–4 weeks, then annually. *Note: Vaccinate only in areas where infection is likely.*

MEDICAL SUPPLY LIST

4-101. In addition to the supplies contained in the veterinary first-aid kit, Table 4-4 lists items that are consolidated into a single kit for use in more definitive care.

Table 4-4. Consolidated List of Items Used for Definitive Care

Item	NSN
Kit, minor, surgical	NSN 6545-00-957-7650
Float, dental, veterinary	NSN 6515-00-938-4301
Needle, hypodermic 20-gram (g), 1 1/2-inch	NSN 6515-01-003-2368
Needle, hypodermic 18g, 1 1/2-inch	NSN 6515-00-754-2834
Needle, hypodermic 25g, 1-inch	NSN 6515-01-037-5590
Basic IV drip set, 10 or 15 drops/milliliters (ml)	NSN 6515-01-332-1276
Pump, injection and suction, veterinary	NSN 6515-00-938-4718
Suture, nonabsorbable size 3-0, with cutting needle	NSN 6515-00-054-7444
Suture, nonabsorbable size 0, with cutting needle	NSN 6515-01-195-7701
Twitch, chain	NSN 3770-00-191-8055
Clipper, hair	NSN 3770-00-804-4700
Stethoscope, bell/diaphragm	NSN 6515-01-304-1027
Thermometer, digital, flexible	NSN 6685-00-444-6500
Tube, nasogastric, veterinary	NSN 6515-01-153-5387
Gloves, examination, veterinary	NSN 8415-01-359-7935
The Merck Veterinary Manual	ISBN 0911910557
Orsini and Divers Manual of Equine Emergencies	ISBN 0712624251
Syringes, 3-ml, 5-ml, 10-ml, and 60-ml Luer tips	Varies per Size

PHARMACOLOGICAL LISTING

4-102. Table 4-5 lists antibiotics, antiparasitics, and antifungals used to treat pack animals. It also lists the recommended drugs, dosages, and routes of each item used.

Table 4-5. Pharmacological Listing

Drug Class: Antibiotics	Drug	Route and Dosage
Penicillins	Sodium	pcn G, IV, 10,000–20,000 immunizing units (IU)/kg, q6 hr
	Potassium	pcn G, po 25,000 IU/kg, q6 hr
	Procaine	pcn G, IM, subcutaneously, 22,000–44,000 IU/kg, q12 hr
Cephalosporins	Cefazolin	IV, IM 20–25 mg/kg, q6 to 8 hr
Aminoglycosides	Gentamicin	IM, 4 mg/kg, q24 hr
Tetracyclines	Oxytetracycline	IV, 5 mg/kg, q12 to 24 hr
Sulfonamides	Trimethoprim-sulfa	20 mg/kg q12 hr
Drug Class: Antiparasitics	**Drug**	**Route and Dosage**
Anthelmintics	Thiabendazole	po 44 mg/kg, q24 hr or 22 mg/kg, q12 hr
	Ivermectin	0.2 mg/kg po once
	Fenbendazole	10 mg/kg po once
	Pyrantel Pamoate	6.6 mg/kg po once
	Oxibendazole	15 mg/kg po once
Drug Class: Antifungals	**Drug**	**Route and Dosage**
	Iodine	Topical
	Copper Sulfate	Topical
	Tolnaftate	Topical
	Miconazole	Topical

EUTHANASIA

4-103. Euthanasia refers to the ending of a life, in a humane manner, to relieve suffering from illness or injury. While caring for animals during combat conditions, the handler may find that serious illness or injury will necessitate the destruction of an animal that is beyond the scope of available medical treatment. Euthanasia under field conditions is frequently a rather brutal affair, without the poisonous gases and injectable venoms used in the clinic. Despite this lack of "civilized" methods of destruction, every effort should be taken to effect euthanasia as rapidly and painlessly as possible. In addition, any personnel near the animal must exercise caution to avoid injury to themselves should the animal be grievously wounded but not immediately killed. The thrashing of these powerful animals will result in injury to bystanders and could create panic among other animals stationed nearby. The handler should keep other animals as far away from the site as possible.

4-104. Equines, despite their immense size and strength, are fairly fragile animals. Nearly any injury resulting in a fracture of the legs will necessitate destruction of the animal. If the animal was otherwise healthy, the meat of the animal may be used as food. The unit should avoid eating animals that were ill or infested with internal parasites, or that were recently treated with medications. Medications have specific meat and milk withdrawal times that specify how long after their use one must wait to consume the animal or its by-products (for example, milk) as food. These withdrawal times are listed in veterinary drug reference books.

4-105. When a veterinarian is not available to the pack animal detachment in the field, the detachment medic (18D) should receive medical training in the use of euthanasic drugs from a large animal veterinarian for a rapid, humane, and silent manner of euthanasia. A variety of methods are available for field euthanasia. Of these, most are impractical because they require a very advanced degree of anatomical expertise or require dosages of common drugs in amounts larger than would logically be available. Consequently, simple mechanical methods of destruction are all that will be available to the average soldier. Personnel should always remember that an animal euthanized with these agents **cannot be used for food**.

Chapter 5

Packing Equipment

This chapter explains some of the most commonly used packing equipment, how it should be maintained, and how it should be fitted to the animal. There are variations in equipment; however, the principles for using it remain the same.

NOTE: Most injuries and wounds to pack animals result from poorly adjusted equipment, such as the saddles, pack frames, thickness of pack pads, and harnesses.

SELECTION OF EQUIPMENT

5-1. Before departing, the detachment should try to determine the size of the animals with which they will be working. Locally available packsaddles, bridles, halters, harnesses, or any other forms of equipment are not always necessarily the best available. The detachment leader should research carefully the species he intends to use, the method he intends to use them with, and how he proposes to get from point A to point B. If buying overseas, the detachment leader can bargain for equipment but should never "buy cheap." Bad equipment will cause difficulties continuously, especially if it breaks down because it was insufficient to begin with. The detachment should be conservative with its funds and invest in quality. If the size of the animals can be determined, the detachment will have an idea of what size to buy. When in doubt of an animal's size, the detachment should get the equipment a little larger and punch some extra holes for adjustment. Equipment that can be adjusted should always be the preferred option. Having equipment that will not fit the animals and cannot be sold or bartered for suitable gear is useless. If local equipment is purchased, the detachment should make sure it fits and, if possible, should always use the same well-fitted equipment on the same animal.

PACKSADDLES

5-2. The type and variety of packsaddles available are extremely wide and are controlled by the type of animals being used and their relative sizes. The types of packsaddles discussed in this chapter are the sawbuck (or crosstree), the Bradshaw, the Decker, and the new hybrid-version sawbuck and Decker saddles. They are all very adaptable to different types of loads and, therefore, are the best type to use in carrying cargo of different weights and sizes. These saddles can carry side loads, top loads, or a combination of the two. All of these saddles can accommodate standard packing hitches with a minimum requirement for tying and threading lash ropes to secure a load. They can be quickly and easily packed and unpacked.

SAWBUCK SADDLE

5-3. The sawbuck saddle is one of the oldest types of packsaddles still in use throughout the world. It is the simplest and most easily constructed. The sawbuck saddle (Figure 5-1) consists of two sidebars connected at the front and rear by crosspieces (bucks) forming an "X" over the spine. The two sidebars are called "humane bars" if they are curved to fit the shape of the animal's body. The humane bars allow the saddle to fit the animal better and make carrying the load more comfortable. Most sawbuck saddles produced today have humane bars. Cargo is carried on the saddle in panniers, pannier bags, mantees, or boxes that are hung from the bucks or carried by hitches and slings.

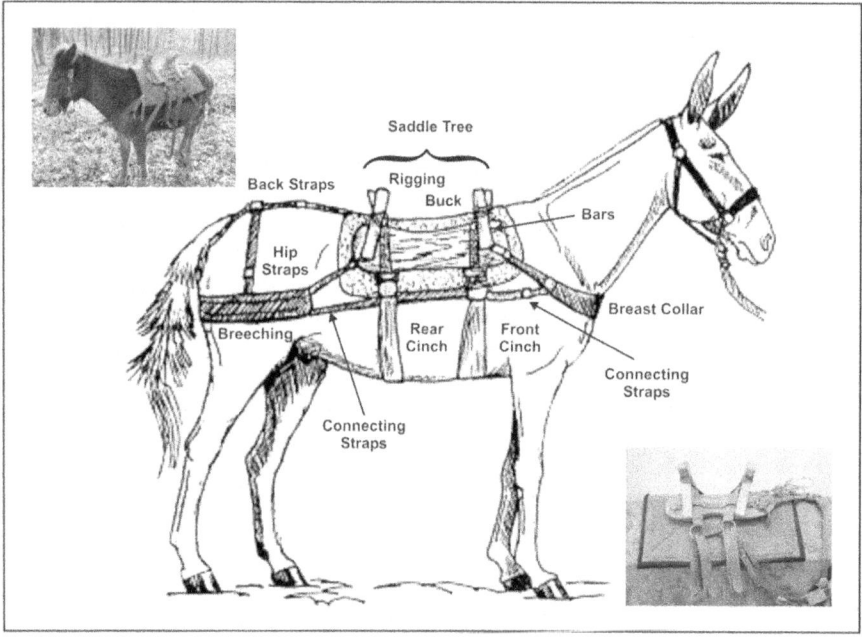

Figure 5-1. Sawbuck Saddle on a Mule

BRADSHAW SADDLE

5-4. The Bradshaw saddle is only slightly different from the sawbuck saddle. The Bradshaw in Figure 5-2, page 5-3, has extended wood crosspieces and a steel bar near the bottom, connecting them on both sides. The extended crosspieces give more protection to the animal's flanks from bigger loads, reducing the weight that is applied directly to the animal.

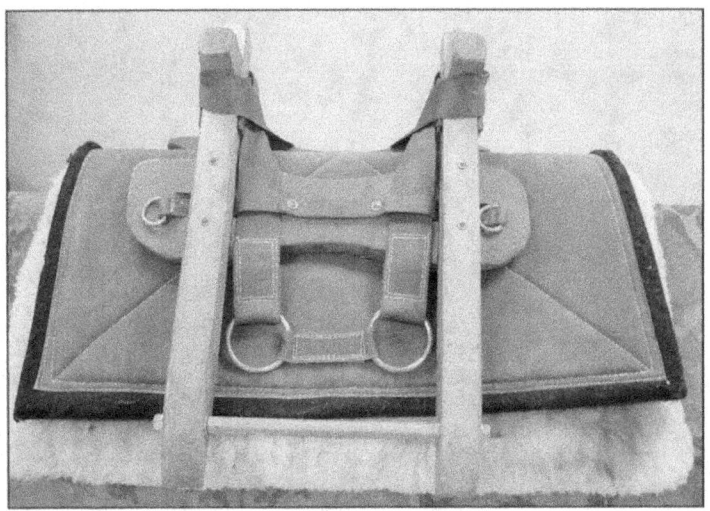

Figure 5-2. Bradshaw Saddle

DECKER SADDLE

5-5. The Decker saddle is made basically the same way as the sawbuck. The difference is that it has metal hoops (arches) instead of cross bucks holding the humane bars together (Figure 5-3, page 5-4). Some packers prefer the Decker saddle to the sawbuck saddle because the humane bars on the Decker saddle can be adjusted by bending the metal hoops to fit the animal better. The animal carries cargo in nearly the same manner as on the sawbuck. The difference with the Decker saddle is that the panniers are hung over the metal hoops instead of the cross bucks. Decker saddles have "ears" welded to the metal hoops to keep the panniers from slipping off or to run the sling rope around when using one. The ears are pieces of metal stock approximately 2 inches long and 1/2 inch in diameter and are welded to the metal hoops near the top on either side. Panniers with adjustable straps are sometimes secured to the Decker saddle by running the strap under the metal hoops and then fastening the buckle.

HYBRID-VERSION SADDLES

5-6. The hybrid-version sawbuck and Decker saddles in Figure 5-4, page 5-4, have fully floating plastic composite humane bars with fully adjusting crosspieces (Bucks) made of aluminum. This system was designed to greatly reduce back sores and provide more energy to the pack animal. This system weighs less than half of the standard sawbuck packsaddle and is rigged the same as the more conventional sawbuck and Decker packsaddles.

FM 3-05.213

Figure 5-3. Decker Saddle With Half-Breed Saddle Cover

Figure 5-4. Hybrid Saddles

SADDLE HARNESS

5-7. A variety of different straps form the harness. It holds and stabilizes the saddle to the animal. The main components of the harness are the cinches, breast collar, breeching, and crupper. The packer secures all of the straps either directly to the saddle or to the rigging.

Rigging

5-8. The rigging is a leather strap that wraps around the cross bucks on the sawbuck and around the metal hoops on a Decker saddle. The packer secures it to the humane bars with screws and allows the ends to hang below the bars. He attaches the rigging rings to the ends of the straps. He then secures the latigos (which are used to secure the cinches) and the connecting straps (which help hold the breeching and breast collar in place) to the rigging rings.

Cinches

5-9. Cinches are the part of the harness that hold the saddle to the animal and provide stability to the saddle once it is packed. Most packsaddles have two cinches, one in front and one in the rear. They fit around the belly or barrel of the animal. The packer secures the cinches to the rigging rings by the latigos.

Breast Collar

5-10. The breast collar provides stability to the saddle while the animal is traveling uphill. It keeps the saddle from moving rearward over the animal's kidneys and rump. The breast collar is usually made of leather or cotton duck material and is approximately 4 inches wide. Two connecting straps from the front of the humane bars secure the breast strap, which is usually made of wool or tail and mane hair.

Breeching

5-11. The breeching fits around the animal's hips and keeps the saddle from moving forward over its withers while traveling downhill. The breeching is made of leather or cotton duck material and is approximately 4 inches wide. It is held in place by connecting straps, back straps, quarter straps, and hip straps. The connecting straps run from the rigging rings to the ends of the breeching. The back straps run from the rear of the humane bars to a metal ring (called a spider) that rests on the animal's rump. The hip straps run from the metal ring to the breeching.

Crupper

5-12. The crupper is a leather strap that runs under the animal's tail and attaches to the metal ring. It keeps the saddle from slipping forward. Cruppers serve the same purpose as the breeching but are not widely used. They can be used in place of, or with, the breeching.

HALTER AND PACKING EQUIPMENT

5-13. A halter is a control device that fits around an animal's head and must be placed on the animal before packing it. By controlling the animal's head, a person can control the animal. The halter is used mainly for leading rather than riding an animal. Without one it would be next to impossible to maintain control of an animal for any length of time. Halters are simple devices constructed generally of nylon webbing or rope. Another essential item for controlling an animal is the lead line or lead rope. A lead line is a piece of rope usually 1/2 or 3/4 inch in diameter, made of pliable material, usually cotton or nylon, and approximately 10 feet long. The line attaches to the chin ring on the halter under the animal's chin. It can be permanently spliced into the chin ring or attached by means of a snap (Figure 5-5, page 5-6).

NOTE: A person should never wrap the lead line around his hand. If the animal is frightened, the person could be entangled in the rope. A person should S-fold the rope and place it in the center of his hand.

Figure 5-5. Halter With Lead Line

HALTER PLACEMENT

5-14. To place a halter on an animal, the handler should first gain control of the animal (Figure 5-6, page 5-7). He places the halter and lead line over his left shoulder so the animal does not become familiar with seeing him walking into the pasture with the lead line in hand. He approaches the animal talking to it or by offering the animal grain.

SADDLE PACK PAD

5-15. The saddle pack pad provides the only protection between the loaded saddle and the animal's hide. Pack pads are usually thicker than riding pads because a pack animal carries "dead" weight. A riding animal carries "live" weight, which means that a rider will shift his weight as the animal traverses different types of terrain. A pack load is dead weight, which means it will not move with the animal. It is important that the saddle pack pad be made of a material that will stay soft and not compress and get hard during use. Pack pads are usually 30 x 30 inches, and made of fiber, fleece, felt, or hair with a canvas portion on one side to protect the animal's back from the bars and equipment on the packsaddle.

Once the handler determines the animal is calm, he reaches around the animal's neck with his left hand holding the end of the lead line.

The handler then grabs the lead line with his right hand. He should now have control of the animal and the animal should follow his commands.

He holds the halter with the lead line ring down, with the closed end to the front of the animal's nose, and the open end to the rear of the animal's head.

The handler slips the closed end over the nose, brings the running end of the open end over the back of the animal's head behind the ears, and attaches it. He keeps control of the lead line during this process by holding the lead line with one hand or by placing one foot on the running end of the lead line to prevent the animal from pulling the halter away as he puts it on.

The handler now checks to see if there are any twists in the halter straps and that the halter is properly adjusted. If the halter is properly adjusted, the molar ring should be about 2 inches from the molar bone, the throat latch should be a hands width between the animal's jaw and throat latch, and the noseband should be two fingers' width from the animal's nose.

Figure 5-6. Placement of Halter on Animal

NOTE: The handler should not connect the lead line to the molar or cheek ring. It should only be connected to the chin ring.

5-16. Some packers use what is called a cheater pad on animals that have high withers or very thin backs, or for extra protection. These small pads are usually 30 x 30 inches, and made of fiber, fleece, felt, wool, or hair (Figure 5-7). The cheaters can be placed between two regular pack pads in the withers area to raise the front of the packsaddle off the withers and yet have the saddle sit evenly on the pack animal's back. It is important that the cheater pad is the first pad placed on the animal's back. Not all animals will need this extra padding.

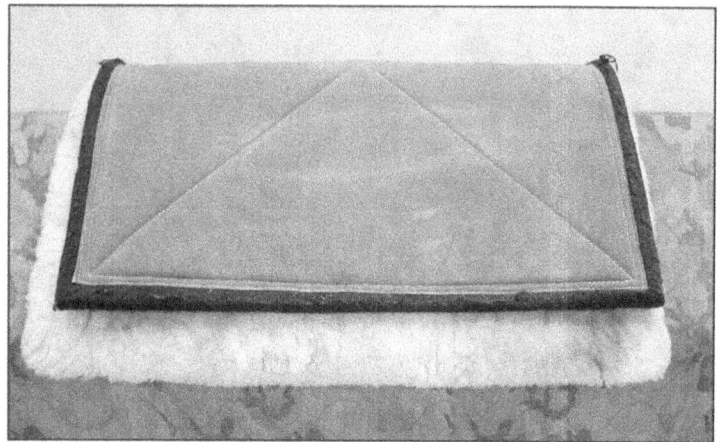

Figure 5-7. Pack Pad With Cheater Pad

SADDLE COVER

5-17. The saddle cover, sometimes called a half-breed, is used when it is necessary to protect the animal's flanks from the load or the saddle rigging. The saddle cover is commonly made of two pieces of canvas sewn together with slots cut in it so that it will fit over the cross bucks or metal hoops. The saddle cover can have padding sewn into it or it can have stuffing slits so that the user can fill it with as much padding as is necessary for the load. It usually has sideboards made of wood, 2 to 4 inches wide, attached near the bottom to provide further protection to the animal's flanks. The saddle cover can be placed over the saddle once it is on the animal, or it can be placed over the humane bars and under the rigging (Figure 5-8, page 5-9).

MANTEE

5-18. A mantee is a tarp usually made of canvas that is lashed over the top load secured on a packsaddle or used to wrap up cargo that is to be placed or slung on a packsaddle. A mantee protects the load from trees, bushes, rocks, or anything else the animal may brush against while it is moving. It also helps protect the load from rain and snow. The size is generally 8 x 8 feet. Each mantee has an assigned cargo rope that goes with it. Mantees are folded four to a bundle with ropes located in the center. However, a packer can use any size that will protect the load.

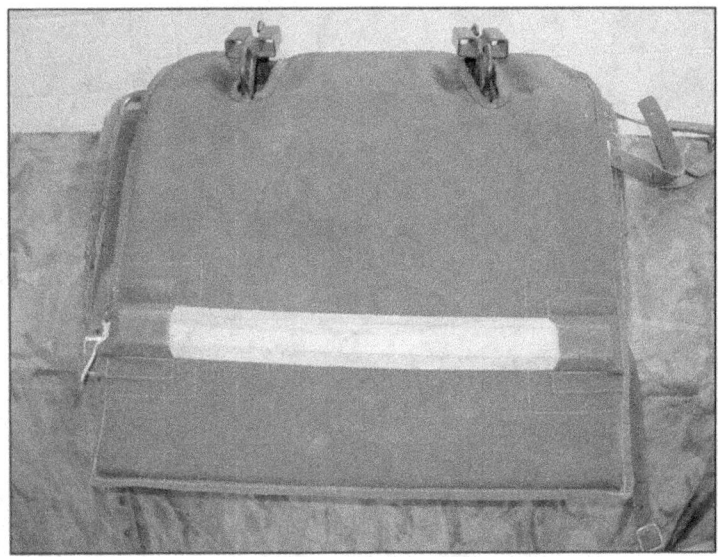

Figure 5-8. Saddle Cover

LASH ROPE AND CINCH

5-19. The lash rope and cinch (Figure 5-9, page 5-10) secure the load and mantee to the animal by means of different types of packing hitches. The lash rope is usually 36 feet long, 1/2 to 5/8 inch in diameter, and made of manila or nylon. The lash cinch is usually made of heavy cotton duck material with a ring at one end and a hook at the other. The lash rope is spliced to the cinch at the end with the ring.

SLING ROPE

5-20. The sling rope fastens to the front cross buck on a sawbuck saddle or the front and rear arch on a Decker saddle. It secures loads that will not fit in panniers or secures panniers that do not have straps. A sling rope is normally 36 feet long, 1/2 inch in diameter, and made of hemp, nylon, or polyester.

PANNIERS

5-21. Panniers are cargo containers that hang from the cross bucks on a sawbuck or the metal hoops on a Decker saddle (Figure 5-10, page 5-10). Some panniers do not have straps to hold them to the saddle. If they do not have straps, the packer can use the sling rope to hold the panniers in place. The dimensions of most commercially made panniers are approximately 26 inches long, 19 inches high, and 12 inches deep. Panniers are made of many different types of materials. Some of the most popular are canvas, aluminum, hard plastic, plywood covered with fiberglass, or a wood frame covered with cowhide.

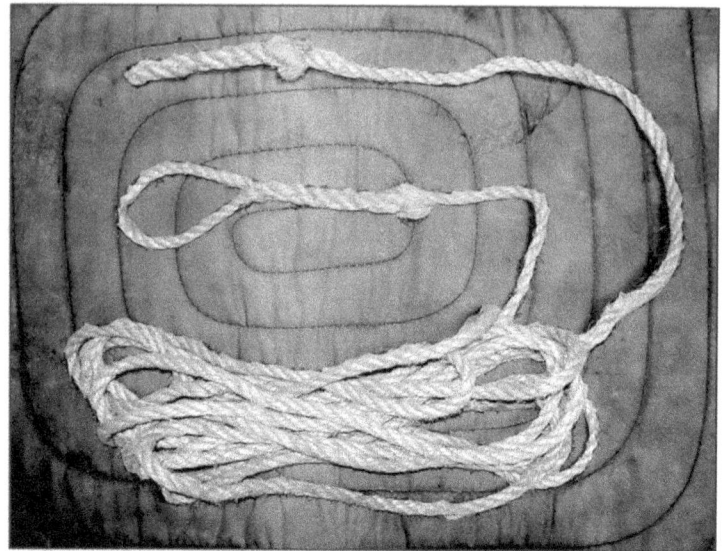

Figure 5-9. Lash Rope and Cinch

Figure 5-10. Two Types of Panniers

CARE OF EQUIPMENT

5-22. The packer has the primary responsibility for the care and preservation of the packsaddle and equipment. The packer should perform the routine cleaning, preservation, and daily inspection for the packing equipment to ensure mission readiness.

5-23. The packer checks saddles daily to ensure there are no cracked, broken, or loose parts. He checks the cross bucks or hoops, humane bars, rigging, and latigos. He tightens any loose items and repairs or replaces cracked or broken parts. The packer cleans any mud or other debris from the saddles that may have collected during the day's movement. He should check all parts of the harness for signs of wear, breaks, cleanliness, and serviceability (Figure 5-11).

Cinches	• Check cinches for any broken strands. • Check the cinch rings for any signs of cracking or metal fatigue. • Repair or replace any damaged pieces.
Breast Collar	• Check the breast collar and connecting straps for any signs of wear or debris collected during the day's movement. • Check the fastening devices to ensure serviceability. Repair or replace items as necessary.
Breeching	• Check the breeching and all associated straps for cleanliness and serviceability. • Pay close attention to all fastening devices. • Repair or replace items as necessary.
Crupper	• Check the crupper (if used) for cleanliness and serviceability.

Figure 5-11. Checks for Harness Parts

5-24. The packer brushes or shakes out the saddle pads after he removes them from the animal. He checks them thoroughly for any foreign objects that can come in contact with the animal's hide. The packer then lays the saddle pads where they can air out and dry.

5-25. The saddle cover should be treated in the same manner as the saddle pad. In addition, the packer checks the wooden bars for serviceability. He checks the padding to see if some should be added or taken out. He also checks the mantee for rips, holes, and general serviceability.

5-26. The lash and sling rope should be checked for signs of wear. The packer replaces or repairs them immediately if there are signs of excessive wear. Breaking a lash or sling rope on the trail could cause a wreck. The ropes should be kept as dry as possible, especially if the rope is made of hemp. Wet weather causes hemp ropes to become hard to manage. Wet hemp ropes will dry and stretch out, leaving the pack loose or uneven. The packer should never throw the ropes on the wet ground while packing or unpacking an animal. They should be hung on a tree limb or laid on a tarp or any dry surface until ready for use.

5-27. The packer checks the lash cinch often for signs of the material fraying. Badly frayed material can tear on the trail and cause a load to come loose. Frayed material can also rub on the animal's belly and cause a sore. The packer also checks the cinch ring for signs of wear and metal fatigue, replacing or repairing items as necessary.

5-28. The packer cleans panniers and checks them for overall serviceability. Most importantly, he checks the straps and fasteners on the straps that hold them to the saddle.

5-29. All leather items should be kept clean and free of grit and dirt. In cleaning off mud or excessive dirt, the packer uses a grooming brush or a blunt piece of wood and a sponge, lukewarm water, and saddle soap. After cleaning, the packer applies neat's-foot oil to the leather for both protection and appearance. He also cleans all metal parts as well.

5-30. Equipment made from canvas and duck material should always be kept free of dirt and mildew. A frequent brushing will remove dirt. The best remedy for mildew is air and sunlight.

FITTING AND ADJUSTING THE SADDLE

5-31. The proper positioning of the saddle and correct cinch adjustment are very important factors. Improper adjustment may cause injury to the animal or may affect the time and distance the animal can carry its load. Before placing anything on the animal's back, it is important to be sure that the animal feels comfortable about it. Many will get skittish if they are unfamiliar with the equipment. A good way to get the animal accustomed to the equipment is to hold it in front of the animal and let him see it and smell it. The animal should be saddled with the packer standing on its left (onside). When making adjustments to the saddle or harness, the packer will have to move to the right (offside) of the animal. The packer should always move around the rear of the animal when moving from one side to the other while saddling. To do this, he gets close to the animal and maintains contact with it by keeping a hand on its rump. If the animal should kick while he is moving around it, receiving a kick from a short distance is much better than from a long distance after the full force of the kick has been generated.

GROOMING

5-32. The handler should always groom the pack animal before saddling. While grooming, the handler removes any debris from the animal that may cause saddle sores. He also checks for any sores from the previous movement or sores that may have occurred during the night. The handler treats the sores as necessary or asks the medic or vet for assistance. If there is any question as to the ability of the animal to carry a load, the handler should ask the medic or vet to make a determination. The animal should not aggravate existing sores.

PAD PLACEMENT

5-33. The handler checks the saddle pad thoroughly for any foreign objects before placing it on the animal. He places the saddle pad square on the animal's back, forward of where it needs to rest, and then slides it rearward into position. Sliding the pad rearward will make the hair lie naturally and prevent sores. The forward portion of the pad should be over the withers with its forward edge about a hand's breadth in front of the rear edge of the shoulder blade.

5-34. The animal may need more than one pack pad or cheater pad depending on the shape of its back. The best way to determine how many pads the animal needs is to set the packsaddle on him, and then check the clearance between the saddle and his withers. The handler should allow for the saddle settling down on the pads after the animal is loaded. If there is any chance of the saddle forks coming in contact with the top of the withers, the handler can put another pack pad under the saddle. He should be careful not to get too much padding on the withers so the packsaddle is pinching the withers from the pads being too thick. The purpose of putting extra padding on a high-withered animal is to raise the packsaddle off the tops of the withers. If the pads being used are the same thickness throughout, the handler will not gain much. The pad must be thicker over the withers and thinner toward the rear of the animal. A cheater pad sits on the withers only and that is why it is usually the best thing to use for a high-withered animal. Also, pads always have a tendency to slip to the rear of the animal.

SADDLE PLACEMENT

5-35. The handler places the saddle square on the pack pads allowing 4 to 6 inches of padding to be exposed in the front and rear of the saddle. The forward edge of the saddle will be approximately 2 to 3 inches to the rear of the shoulder blades. It is important to ensure there is adequate saddle pad forward of the saddle to protect the animal's back from the leading edge of the saddle.

5-36. The handler fastens the cinches, the front one first, making sure there are no twists in the offside latigos or the cinches themselves. The front cinch should be a hand's breadth to the rear of the front leg.

5-37. The cinches should be tightened just enough to hold the saddle in place until the breast collar and breeching are adjusted. These adjustments are explained in detail below.

BREAST COLLAR

5-38. The breast collar should ride just above the point of the animal's shoulder and go around the breast below the animal's neck. It should not be tight; its only function is to keep the saddle from slipping back. The handler adjusts the breast collar so that it is snug when a front leg is fully extended. He makes sure the two connecting straps are adjusted to the same length so that the breast collar rides evenly.

BREECHING

5-39. Once the saddle is in place and cinched lightly, the handler adjusts the breeching. The handler—

- Places the spider on the animal's rump and adjusts the back straps so that it rests on the highest point of the animal's shoulder. The handler ensures that both straps are the same length so the ring will stay centered.
- Adjusts the hip straps so the breeching rides approximately halfway between the base of the tail and the bottom of the hindquarters. He

adjusts the straps on both sides to the same length to ensure the breeching rides level.
- Attaches the crupper if used. He lifts the animal's tail and slides the crupper under it. He must always be very careful when working around the animal's hind end. Some animals will try to kick.

NOTE: The best thing for the handler to do is stay very close to the animal with his side touching the animal's flank. The animal will telegraph any intention to kick by tightening its muscles. The handler should watch the animal's ears. If it intends to kick, it will lay its ears back.

- Talks to the animal reassuringly while lifting the tail to slide the crupper under it. The crupper should be snug, but not tight, against the base of the tail but not tight. The handler ensures both connecting straps on the crupper are the same length for even riding.
- Adjusts the connecting straps. A good rule of thumb to follow for adjusting the connecting straps is to make sure that, when the animal is walking and one hind leg reaches its rearmost position in the stride, the breech strap is firmly against the animal's rump. The breeching should not be so tight as to hinder the animal's natural gait. However, if it is too loose, it is useless.
- Connects the quarter straps to the front cinch ring. The handler adjusts them snugly to keep the front cinch from moving forward while the animal is moving. He makes sure the upper quarter strap is snug and the lower quarter strap has a 1-inch sag to it.
- Completes initial adjustment of the breeching and then pulls the animal's tail out from under the breeching. The crupper adjustment discussed above explains the best method.

5-40. The breeching is now adjusted approximately to where it should be. The handler walks the animal around to check how the breeching is riding on the animal and makes adjustments as necessary. When traveling down long steep slopes, it may be necessary to tighten the back and connecting straps some to keep the saddle from slipping over the withers and shoulders.

CINCHING

5-41. Proper cinching is essential because the packsaddle covers so great an area of moving surface. Excessive binding of the front cinch may injure the back and sides, interfere with breathing, or cause cinch sores. The handler uses the front cinch to secure the saddle in place and makes sure it is tighter than the rear cinch. The rear cinch keeps the saddle from rocking front to rear as the animal is walking; therefore, the rear cinch does not need to be as tight as the front cinch. Since the hind legs are the propelling members, the hindquarters move from side to side and up and down. These movements should not be restricted by cinch pressure. There must be no interference with the animal's locomotion. Only experience can teach a person how to determine the exact amount of cinch pressure needed. A safe rule to follow is to give the front cinch sufficient pressure to hold the saddle in place. Usually, one finger should pass easily between the front cinch and the animal's chest.

5-42. The rear cinch should be tight enough to limit the rocking motions of the saddle and help prevent the saddle from slipping forward. The rear cinch should not be so tight that the whole hand cannot be slipped under it. In testing cinch pressure, the handler should be able to insert a finger from the rear to the front so that, when it is withdrawn, the hair does not ruffle. Ruffled hair may cause sores. Excessive binding of the rear cinch will cause a pack animal to tire quickly. It is very important to center the cinch on the animal's belly. The cinch rings on either side of the cinch should be the same distance from the rigging when the cinch is pulled tight. This type of fit is called a ring-to-ring check. An uneven cinch could cause cinch sores or cause the saddle to slip. The latigos should be secured on both sides of the saddle with a quick-release knot. This knot is important because it allows the load to be released quickly from an animal if the load should fall onto its side or upside down. With other knots it is necessary to cut the load away from the animal. Having to cut the load may cause injury to the animal and will ruin the latigos or cinches on the saddle.

5-43. To fasten the cinch on the animal, the handler runs the latigo through the cinch ring (the end of the latigo is run through the cinch ring so that it comes through the ring toward the packer) and back up through the rigging ring (the end of the latigo is run through the rigging ring so that it goes through the ring toward the animal). If the latigos are long, or the animal has a small barrel, they may have to be wrapped more than once to take up the excess length of the latigo. The knot used is a "half-Windsor." To form this knot, the handler brings the running end of the latigo around the portion of the latigo running through the rigging ring, up through the rear of the rigging ring, and down through the loop just formed. He forms the quick release by passing the running end of the latigo up through the knot just formed (Figure 5-12).

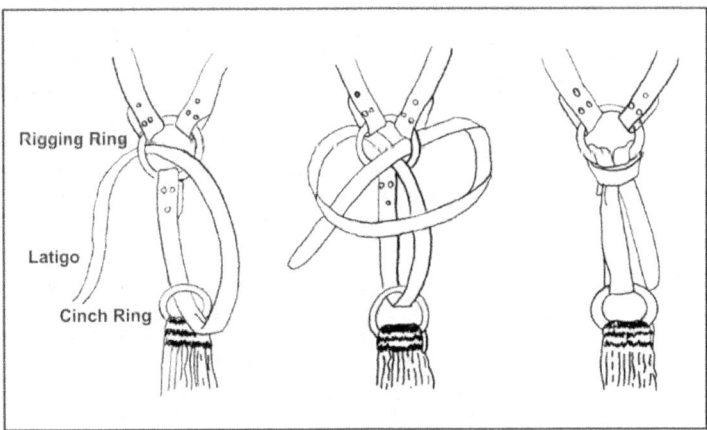

Figure 5-12. Latigos Fasten the Cinches to the Rigging Ring

FINAL ADJUSTMENTS

5-44. When the saddle is in the right position on the animal and the breast collar, breeching, and cinches are snug, it is time to "untrack" the animal before packing it. Many times an animal will force air into its lungs and belly when it realizes that it is going to be saddled. The animal does this to make the saddle fit more comfortably. Once the animal starts moving, it will expel the air and the saddle will fit more loosely. If the animal is packed without untracking it, the saddle could slip shortly after movement begins and the whole load will have to be repacked. To untrack an animal, the handler takes the animal from the place it was saddled and walks it around for approximately 30 seconds. He ties the animal up and tightens the cinches again if needed. The animal is now ready to be packed.

MARKING SADDLES

5-45. After fitting the saddle to the pack animal, the handler marks it with the animal's name or number and uses it with the same animal throughout the movement. The same saddle pads and, if needed, cheater pads should be kept with the animal so the saddle fits the same every time. This technique will save time refitting the packsaddles every time the unit prepares to move.

UNSADDLING THE ANIMAL

5-46. The animal should be unsaddled in the opposite sequence that it is saddled. By following this sequence, the saddle will be stored in a manner that will make saddling the animal quick and easy the next time it is needed. Since the breast strap, breeching, and cinches were adjusted properly when the animal was saddled, the rider can maintain a proper fit if he loosens only the quarter straps, cinches, and breast collar (onside strap and the strap going to the hobble ring). To properly unsaddle an animal, the handler or rider should—

- Unfasten the breast collar strap from the front hobble ring on the front cinch and from the onside connection. Take the breast collar strap from around the neck of the animal and run the onside breast collar strap under the bucks from the rear to the front leaving the buckle visible at the rear of the saddle.
- Lay the excess strap on the animal's neck and fold the excess breast collar into the center of the saddle between the bucks.
- Unfasten the quarter straps and hook them onto the rear rigging rings and unfasten the rear cinch.
- Put the running end of the latigo through the rigging ring twice and bring the running end around the latigo, up through the rear of the rigging ring and down through the loop just formed. He should repeat the procedure for the front cinch.
- Slide the saddle rearward to loosen the breeching. If a crupper is used, unfasten the onside strap and lift the animal's tail to remove the breeching.
- Lift the breeching over the rump and place the spider and crupper, if used, between the bucks. Lay the breeching across the saddle between

the bucks. Fold the excess straps into the center of the saddle between the bucks and fold the cinches into the center of the saddle on top of the breeching.
- Take in the portion of the breast strap that runs under the bucks and loop it around the breeching and cinches folded into the center of the saddle. Ensure the strap runs diagonally across the saddle so the strap will tighten as much as possible and secure the strap with the buckle.
- Store the saddle appropriately and ensure this saddle is used on its designated animal.
- Remove the saddle pads from the animal and shake them out or brush them. Place the pads over the saddle with the side that was against the animal facing out so they will dry. If a sling rope is used, wrap it around the bucks in a figure-eight manner.
- Roll and store the lash rope (if not used as a high line) as follows:
  - Hold the lash cinch by the cinch ring and coil the rope.
  - Grasp the lash cinch by the hook and wrap it around the coiled rope once and place the hook through the cinch ring.
  - Hang it by the hook for storage.

SADDLING WITH A FITTED SADDLE

5-47. Saddling an animal with an already-fitted saddle is much less time-consuming than saddling and having to adjust the breeching. When saddling with a fitted saddle, the handler or rider—

- Properly grooms the animal.
- Places the saddle pad on the animal.
- Places the saddle on the animal correctly.
- Loosens the strap holding the cinches and rigging together between the bucks.
- Lets the cinches fall to the offside of the animal.
- Pulls the breeching from the stowed position and fits it around the rump of the animal.
- Ensures the saddle is in its proper position.
- Attaches the front cinch loosely.
- Attaches the rear cinch.
- Attaches the quarter straps to the front cinch ring.
- Attaches the crupper, if used.
- Attaches the breast collar.
- Tightens the front cinch.

Since the saddle has already been adjusted to the animal, it should fit properly without further adjustment. However, the handler or rider should check the fit of the saddle and all the rigging to make sure it does fit properly.

Chapter 6

Horsemanship

This chapter discusses horsemanship and should help guide unfamiliar personnel. In many cases, authorities vary on how to perform many functions of good horsemanship, even when presented with the same task. Tactical, environmental, social, and other factors influence how animals are used in a combat environment. The descriptions of equipment and techniques are based on the American-Western style of riding. This style of riding is most familiar to U.S. Soldiers and the most easily adaptable to sustained combat operations.

The emphasis on American-Western style is one of functionality and stability, but as in any style of riding, success is dependent on rider performance. As in most cases, any basic flaws in technique can most likely magnify during combat operations. A pack animal unit will have to improvise equipment or adapt to indigenous equipment in many cases. A basic knowledge of animals, equipment design and function, and tactics should be sufficient for a detachment to perform their mission given any set of circumstances. Reading this chapter is in no way any substitute for experience.

EQUIPMENT

6-1. The following equipment descriptions and instructions for use come from the American-Western style of equipment. In the horse industry, these items are known as tack. Variations of these items are commonplace, but the principles of their design and use remain constant.

BRIDLE AND REINS

6-2. The rider uses the bridle to control the animal when he rides. The bridle consists of various lengths of leather or nylon that the rider can adjust to fit the animal (Figure 6-1, page 6-2). The bridle's basic components include the following:

- The **bit** rests against the back of an animal's mouth and controls the animal by transferring pressure from the reins to the animal's mouth. The rider uses the **reins** or steering lines to command the animal. They are generally leather, approximately 72 to 84 inches long, and can be either split reins (not joined at the ends) or joined at the end. They are attached to the bit at the side rings. The curb chain or strap gives the animal pressure on the bottom of the jawbone when the reins are pulled and assists in stopping the animal. **Cheek straps** run the length of the bridle. Their purpose is to join the bit and the headpiece.

- The **headpiece** runs behind the ears and gives long axis anchor to the bridle. The **browband** runs across the forehead and holds the headpiece in place. The **throatlash** runs from the junction of the headpiece and the browband on each side and under the animal's throat. It further anchors the headpiece. The **noseband** fits around the animal's nose several inches behind the mouth. It keeps the animal from opening its mouth too wide and also provides stability for the bridle.

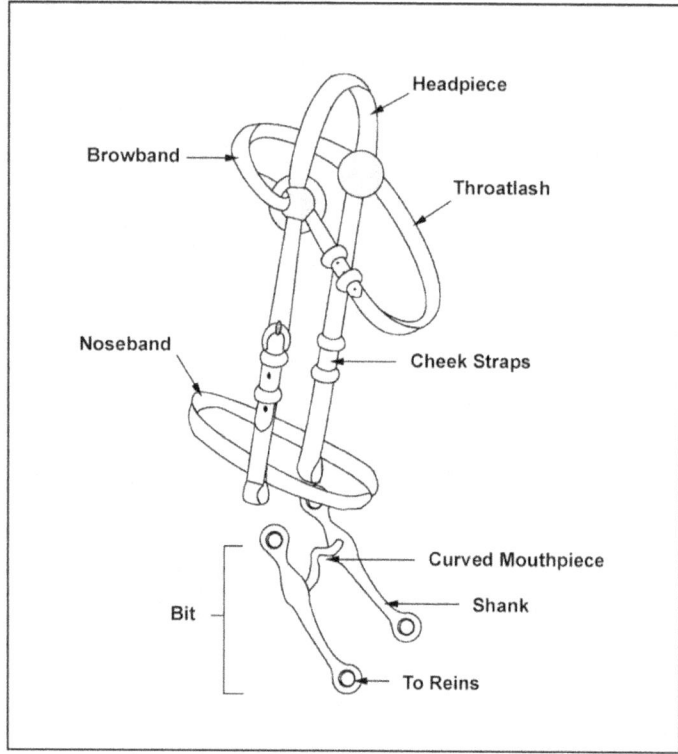

Figure 6-1. Typical Horse Bridle and Standard Bit

BRIDLE AND REINS INSPECTION

6-3. The rider checks all the leather in the bridle, using the same flexing and twisting technique that he used on the billets and stirrup leathers. He pays extra attention to areas of strain, such as where the rein wraps around the bit. Any cracking or separation in that area is a danger signal telling the rider that it is time to invest in a new set of reins.

6-4. The rider should also check all the stitching, especially on the reins. If he has the tools, he may be able to make simple repairs himself. When in doubt, the rider can consult the local saddler for repair work.

6-5. The bit should be checked for rough edges that will damage the horse's delicate lips. The rider replaces any bit that shows signs of roughness and wear immediately.

6-6. As the rider goes over the bridle, he checks the buckles to make sure they are not rusting through and that the tongues are not bent. Bent tongues may allow the buckle to work undone in use.

NOTE: The same safety checks that a rider makes on his bridle should be made on any other equipment that he uses, be it martingales, breastplates, or a complete harness for driving horses. Keeping the leatherwork clean and in good shape will lengthen its useful life span and will be more comfortable for the horse.

6-7. There are many types of bridles. The components identified are not necessarily present on all bridles. Placing the bridle on the animal requires the rider to perform sequential steps (Figure 6-2).

---

1. Stand on the onside of the animal, untie the animal, and remove the halter. Place the reins behind the animal's head and drape them over its neck or the rider's arm.
2. Unbuckle the throatlash.
3. Hold the headpiece in the right hand.
4. Hold the bit in the left hand with the thumb pointed up the axis of the animal's head.
5. Place the right hand on the animal's head between the ears to keep it down.
6. With the left hand, open the animal's mouth at the corner by putting a thumb between its canine and back teeth (the taste of the rider's thumb will cause the animal to open its mouth). Then slide the bit in by pulling with the headpiece in the right hand.
7. Slide the headpiece behind the ears, and place the browband on the forehead.
8. Buckle the throatlash loosely.
9. The bit should be adjusted so that the back of the animal's mouth is drawn up into a slight "smile."
10. Pull loose any restricted mane that is caught under the bridle straps.

Figure 6-2. Steps for Putting the Bridle on the Animal

SADDLE PAD

6-8. Riding pads are generally the same as the pads used on packing animals. The average size is 30 x 30 inches. Chapter 5 explains the use and care of saddle pads.

CHEATER PADS

6-9. Cheater pads can be used on riding animals as on packing animals. Chapter 5, page 5-8, also explains using cheater pads.

WESTERN AND MCCLELLAN SADDLES

6-10. There are a number of saddles on the market today specifically designed for endurance and rough riding. So which one is best for the mission? There is no easy answer. First of all the saddle must fit the horse, and every horse is different. Weight might or might not be an important consideration for the rider, but no packer's saddle should weigh over 30 pounds.

WESTERN SADDLE

6-11. The modern Western saddle is a direct descendant of the deep-seat saddle brought to the Americas by the Spanish Conquistadors in the sixteenth century. It is heavy, and most western saddles put the rider's weight too far to the rear, leading to early fatigue and soreness. Western saddles are designed to keep the rider in place while working cattle, not to provide balance and comfort for miles and miles of rough, mountainous terrain riding. Another flaw of the western saddle is that the rigging is too far forward, contributing to girth soreness or galls. The stiff fenders and stirrup leathers of the western saddle might wear like iron, but can rub a rider's legs raw. Among the characteristics common to Western saddles are the deep seat, saddle horn, long stirrups, and high cantle (Figure 6-3). It also comes with a complement of tie-down straps to secure personal equipment to the saddle.

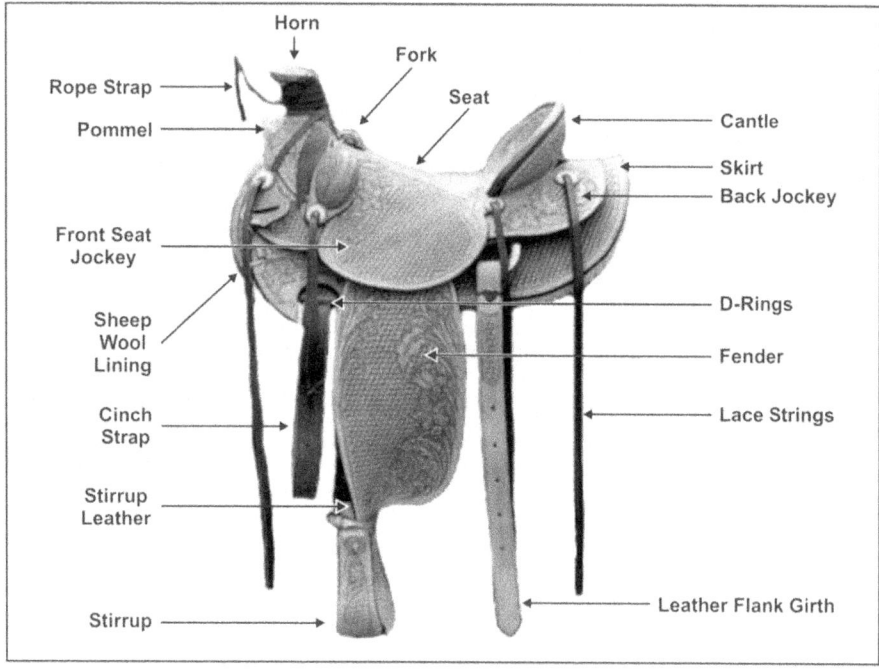

Figure 6-3. Western Saddle

6-12. Western saddles are constructed of wood or synthetics for the tree (frame) and stirrups and covered in leather. If the manufacturer has placed padding in the bars of the saddle, the rider still needs a saddle pad.

6-13. The typical Western saddle will have one cinch to secure the saddle to the animal and may or may not have a flank cinch. Figure 6-4 explains the steps for saddling the animal.

---

1. Ensure the animal is properly groomed.
2. Place the saddle pad on the animal.
3. Ensure the saddle is properly "rolled" for placement on the animal. Being rolled means that the cinch, offside stirrup, and any tie-down straps are pulled over the seat of the saddle.
4. Grasp the saddle by the front center under the horn and the rear center.
   NOTE: If the animal is skittish, or if it does not know the handler, allow it to see and smell the saddle at this time. Make sure the animal is never surprised.
5. From the onside, set the saddle on the animal's back. The front edge of the saddle should be 1 to 2 inches from the front edge of the pad and above the rear of the withers. Grasp the front edge of the pad, and lift it up into the tree to allow space for air to circulate under the pad.
6. Pull down the cinch and connect. Initially, snug the cinch tight enough to secure it to the animal.
7. Connect the flank (rear) cinch, if one is present. It should rest just against the animal's flank and be secured snugly. Connect it to the front cinch in the center with the connecting strap.
8. Lower the stirrups into place. Never drop them against the animal's side.
9. Ensure that no saddle strings or any other objects are between the saddle and the animal.
10. Walk the animal around for a short distance and then adjust the cinches again. The main cinch should be tight enough that three fingers can slide underneath it without much effort.
11. After riding or waiting for a time, recheck the cinches. Exertion or excitement may cause the animal's girth size to change.

Figure 6-4. Steps for Saddling the Animal

MCCLELLAN SADDLE

6-14. The McClellan saddle (Figure 6-5, page 6-6) was adopted by the U.S. War Department in 1859. It was an excellent saddle and remained the standard issue with slight improvements for the remaining history of the horse cavalry. McClellan saddles have a lot to offer the distance rider, but comfort is not one of them. This saddle was used by the cavalry, and was designed to be good for the horse with no compromises toward rider comfort. If a rider finds this saddle satisfactory for his body, he might not find a better saddle for keeping himself in good condition on long rides.

6-15. The typical McClellan saddle features an open, metal-reinforced wooden tree. Saddle skirts of harness leather are screwed to the sidebars. Stirrups are hickory or oak.

6-16. The McClellan saddle can also be used as a packsaddle with no modifications. The McClellan saddle can be rigged in the same configuration as the sawbuck saddle or Decker saddle by using a sling rope to construct a barrel hitch or basket hitch.

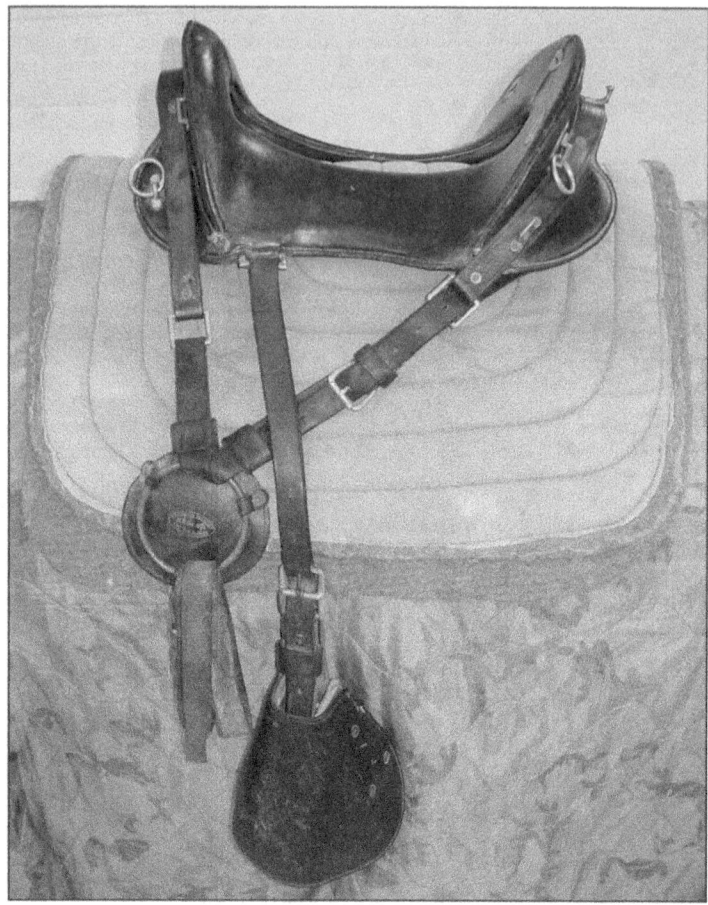

Figure 6-5. McClellan Saddle

6-17. English saddles are lighter and put the rider's weight more forward than Western saddles. They offer closer contact with the horse than Western saddles, but riders sometimes feel less secure in them. The biggest shortcoming of English tack is a lack of distribution of the rider's weight over a large enough area of the horse's back.

RIDER'S EQUIPMENT

6-18. The problem encountered with equipping a U.S. Soldier for mounted operations is that no consideration has been given to the requirements for this type of operation in quite some time. This time lapse can pose various challenges in each mission.

6-19. Many items that are critical for dismounted use are critical to use while mounted as well, but may not perform suitably in both situations. Boots are a prime example. Boots suited for dismounted operations may be detrimental to mounted operations. The very nature of combat operations demands interoperability. This section will address such difficulties. The commander who assigns a unit to perform mounted operations should properly equip those units.

Boots

6-20. Their leather construction, pull-on design, high one-piece uppers, smooth sole, pointed toes, and high heels characterize boots ("cowboy boots") normally associated with American-Western style riding. Boots normally associated with dismounted operations have treated multipiece leather uppers, a lace-up design, lug soles, broad toes, short heels, and are generally shorter than riding boots. Soles that provide traction and footing on the ground can be dangerous when trying to dismount a horse. Lug soles tend to catch on stirrups. Boots with buckles or those with hooks used for speed lacing are not good because the buckles or hooks can catch on the saddle. Boots designed for riding are totally unsuitable for carrying loads over any irregular terrain while walking or for walking any appreciable distance. Riding boots are more difficult to fit and break in as well. Standard military boots, in most cases, will not accept spurs. Standard riding boots will not accept mountaineering equipment (for example, crampons, snowshoes, and skis). Modifying stirrups is not advisable or simple to accomplish. The more suitable approach would be to use military boots with the minimum amount of lug required, shaving down the edges of the soles to prevent the stirrups from becoming wedged into the lug. The widest stirrups available should be used. When using military boots, the rider should exercise care when removing the foot from the stirrup and pay particular attention to how a low heel will affect his ability to maintain the correct seat.

Spurs

6-21. Spurs are removable metal devices that attach to the heel of a rider's boots and assist in the use of his legs as riding aids. There are hundreds of variations of spurs but two basic designs. The classic or cavalry type is a short one-piece spur. The Western style is the star or wheel shape that rotates on a pin. Spurs are not essential for riding but can be useful in controlling an animal. Experience, tactical considerations, the animal's training, and availability of spurs will influence the choice of using them or not. Inexperienced riders may tend to use spurs improperly, which can cause more problems than are corrected. Inexperienced animals may not understand what the rider is asking of them when spurs are applied. When using spurs, the rider should never poke an animal with their points. The spurs' purpose is to enable the horse to better feel the commands of the rider's legs. When applying spurs to the animal's flanks, the rider uses the side of the spur and rolls it upward. A Soldier must remember that spurs may not be usable with his boots and will impair him when dismounted.

Chaps

6-22. Chaps are leather leggings worn to protect the legs while riding. Chaps are not required for riding but do provide considerable protection against chafing from the friction between the legs and the stirrup fenders. They also protect the leg against foliage, limbs, and other such items the rider can brush against while riding. Chaps come in two basic designs: shotgun chaps, which are narrow and zip up the side, and open or bat-wing chaps that connect by ties at the side. The shotgun type, if not too tight, provides the best compromise between tactical use considerations and horse-related work. Shotgun chaps are quieter and will not snag as easily when the rider dismounts. When doing farrier-type work, open or bat-wing chaps are preferred if a farrier's apron is unavailable.

Uniform

6-23. Other than the items already addressed, the requirements for a suitable uniform are the same as for conventional dismounted operations. Some additional factors for the rider to remember are as follows:

- Wear leather gloves when leading pack animals. A rider may find it necessary to take the gloves off during saddling, unsaddling, and tasks such as tying knots.
- Wear a hat to protect the head from low limbs. It should not restrict hearing or vision (particularly peripheral). A chinstrap is helpful since wind and obstructions may cause the hat to be lost and recovery from horseback is impossible. A good choice is the jungle hat or watch cap.
- Carry nothing in the rear trouser pockets.

Necessary Bag

6-24. A necessary bag is a small kit carried by a rider to make field repairs to tack or other equipment. It is normally carried in a saddlebag. There is no specific item list, but typical contents are as follows:

- Leather punch or sharp awl.
- Assorted leather or sail maker's needles.
- Beeswax.
- Waxed sail maker's thread.
- Sewing palm.
- Rivets (for leather).
- Leather bootlaces.
- Small bits and pieces of leather.

Saddlebags

6-25. Saddlebags are of particular importance to the Soldier. There are many different styles of saddlebags. Construction is usually of nylon, canvas, or leather. For military uses, heavy nylon is preferable because it is rot-resistant, abrasion-resistant, and easily repaired. Saddlebags are attached to the rear of the saddle and tied down with the saddle strings located to the rear of the cantle. Locally fabricated models can be designed to work with issued load-carrying equipment (LCE), perhaps attaching to the rider's back. Typical dimensions of saddlebags are 11 x 11 inches x 5 inches thick. When used, as with pack loads, saddlebags must be balanced and sharp contents packed away from the animal.

RIDING TECHNIQUES

6-26. As stated previously, this manual discusses the American-Western style of riding. Riding basics for combat applications are no different from those for pleasure riding. What makes a difference is the skill level required of the rider. Combat conditions force the mounted Soldier to be a master of the basic skills of riding if he is expected to accomplish his mission. Riders must conduct actual training and practice on a regular basis for the skills of a mounted Soldier to remain of high enough caliber to conduct combat operations.

PREPARING TO MOUNT

6-27. Once a rider's animal has been prepared to ride (for example, it is properly groomed and all tack is on and properly adjusted), the rider must ensure that he is ready to ride. All equipment he wishes to carry must already be in place on the animal and himself.

MOUNTING

6-28. The rider unties the animal from its hitch and holds both reins in the left hand before mounting. From the onside, he stands just in front of the saddle, faces to the animal's rear, and turns the stirrup around so he can place his left foot into it. With the rider's left foot in the stirrup, he places his right hand on the saddle horn and left hand on top of the animal's neck. He then swings his right leg over the animal's back. He places his right foot in the stirrup when seated.

NOTE: Only when seated properly in the saddle can the stirrup length be judged to be proper or not. If proper, the knee will have a slight bend in it (approximately 20 degrees) and the rider should just be able to see the tip of the toe over the knee. Another way for the rider to tell if he is seated properly is to stand in the stirrups. There should be enough room to place two fingers between the saddle and his crotch. The rider ensures that the stirrup adjustment hardware is secure after making adjustments.

PROPER SEAT AND AIDS

6-29. The proper seat, or how the rider positions himself in the saddle, is very important to proper riding. If the rider does not sit correctly, the animal will become confused with the commands the rider gives and will not have the proper stability it needs to traverse difficult terrain.

6-30. The position of the rider in the saddle and the way he uses that position is referred to as the "aids." Mastering the aids is the single most important function of riding in control (assuming, of course, that the animal has been properly trained). When in the saddle, the position of the rider's body is very important and relates specific functions to the animal.

6-31. The rider uses his legs to create and control the forward motion of the animal and to assist in steering. When at the normal position, the legs should lie against the animal's side with the heel resting just behind the girth. The feet are placed in the stirrup with the rider's weight resting on the balls of his feet. This position remains the same unless the rider must give the animal a kick. The variance comes from the pressure delivered to the side of the animal by squeezing the leg muscles. To increase forward motion, the rider squeezes more in equal amounts to both sides. When turning, he keeps the inside leg at the girth, and the outside leg back. When halting, he applies light pressure with both legs. When applying any aid, the rider ceases the pressure when the animal responds.

6-32. The primary objective of how the rider distributes his **weight** is to remain in contact with the animal and feel its movements so he can move with the animal. The majority of the rider's weight will be in the stirrups; thus the saying, "standing tall in the saddle." The spine should be straight but not stiff. The remaining weight should be distributed evenly in the seat. If the rider is about to change pace, he should press down in the seat with his pelvis momentarily, without leaning forward, to warn the animal. When riding uphill, the rider must lean forward slightly, and when riding downhill, lean slightly to the rear. In either case, the rider must not exaggerate the movement but keep the weight in the seat. These adjustments are made to assist the animal when it must change its center of gravity.

6-33. The primary function of the **hands** as an aid is to control the reins. Western-style riding (and horses trained Western-style) refers to the technique known as "neck-reining." This style allows the reins to be controlled with only one hand. The reins are held in the dominant hand slightly above the saddle horn. The reins should be held with slight pressure, just enough to maintain light contact with the animal's mouth and low on the neck. To turn the animal, the rider simply moves the rein hand in the direction he wishes to turn and should never over-rein. The rider provides just enough pressure so the animal will understand what is expected. Some experts insist that the animal responds to the pressure created by the reins on the neck, while others maintain that it is a combination of the movements that makes the animal react. In any case, all aids (legs, weight, and hands) must be applied simultaneously and with only enough pressure to achieve the desired effect.

PACES

6-34. As with anything, a rider must "learn to walk before he runs." When training a new rider, it must be done in a controlled environment, such as a corral. The rider starts with the easiest pace first and progresses only when he is comfortable with the level at which he is working.

6-35. The rider gets the horse to **start** moving by making light contact with the horse's side by squeezing his legs. The rider must maintain the proper seat and balance. Riding one hour a week, bareback, will improve a rider's balance rapidly. The rider must not allow his weight or rein hand to come too far forward, as is often the tendency. He should keep the shoulders square to the animal always. A rider may hold the saddle horn if he needs to steady himself at first but should **never** use the reins for any purpose other than controlling his mount.

6-36. Unless the animal has been started in a violent manner, he should assume a **walk** when given the command to start out. The rider maintains all aids as described above when riding at the walk. Also, he should be on the lookout for anything that may be affecting the animal, such as a piece of saddle string under the pad. The rider must remember to "stand" in the stirrups, keeping approximately two-thirds of his weight on the balls of his feet.

6-37. To **halt** the horse, the rider closes the lower legs against the animal's side and lightly tightens, not pulls, the reins. When the animal responds, he releases the pressure but not the control. It will take a rider a little time to determine how responsive any given animal is to this or any other command.

6-38. The **jog and lope** paces are two different paces, but are closely related. The aids are the same as for any pace except that the rider's weight shifts slightly forward. In the jog, the horse will bounce quite a bit, and the rider must learn to move with the animal. To initiate the jog from a walk, the rider repeats the steps used to start the animal from a halt into a walk. The lope is a more relaxing pace for the rider. The animal will tend to bounce less, and it will be easier for the rider to maintain the proper balance. The lope is slightly faster than the jog and is initiated from the jog.

6-39. The **gallop** is the fastest pace of a ridden animal and the most dangerous. It is very easy for a rider to lose control of an animal at the gallop if the animal becomes overexcited, which is often the case when an animal is ordered to gallop. When at the gallop, the rider maintains control with the aids as previously described. The rider must transfer his weight from the seat to the stirrups and the knees. Contact should be maintained on the reins.

6-40. The rider uses the **rein back** to make the animal move straight back. It should not be used over any distance. It is always performed from the halt. The rider performs the rein back by using an "elastic" and staggered pull on the reins and positions his weight slightly forward of center.

LEADING A PACK STRING

6-41. When stringing pack animals together, there are many things to consider: which animals get along with each other, which ones will be easy to lead, which ones will be harder to lead, which ones are more experienced than others, and which ones are carrying the heaviest loads. One of the principal advantages of using animals in the era of modern combat is that mounted elements can move large quantities of material in areas not suitable for conventional transport. When using pack animals, it is often preferable to lead them from horseback to take advantage of the animal's speed. Leading pack animals from horseback is not difficult but the rider must observe the following rules:

- It is best for inexperienced packers to start by leading only one pack animal first. As the rider's confidence and experience level grows, he can add more stock to his string.

- In general, the rider puts the pack animals with the heaviest loads toward the front of the string. The farther back the animals are in the string, the more they have to work to keep up.

- The rider places the pack animal that is easiest to lead up front so the animals do not pull on the rider as he heads up the trail. The pack animals that are a little harder to lead can be tied into the sawbuck saddle or Decker saddle by using a small piece of baling twine (40 to 60 pounds) tied in a pigtail knot (Figure 6-6) as a breakaway from the pack animal ahead. If the rider has an animal that continues to break the pigtail, he should double or triple the pigtails being used. For animals that are not equipped with packsaddles, the rider can use a tail knot (Figure 6-7, page 6-13), that is connected to the lead animal's tail with the lead line of the trailing animal.

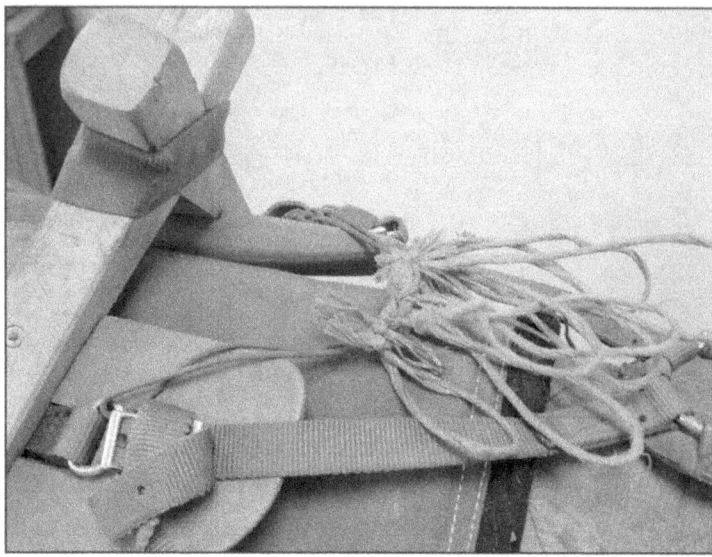

Figure 6-6. Pigtails on Sawbuck Saddle

- For the safety of the rider and animal, there are two ways to tie a pack string together during movement. The choice of tying the animals together is up to the rider and what he is familiar with. The techniques that most mule handlers use are the pigtail and tail knot.
- A pack string should **never** be tied to the lead (ridden) animal. If an accident occurs or the pack string becomes frightened, the rider is in certain danger if he cannot release the string. This point also applies to leading a single animal. The lead line should always be held in one hand. The rider forms a bight, if needed, over the saddle horn, but he must be sure that the lead line can be jettisoned immediately, if required. Likewise, there should be no loose loops in the lead rope because of the danger of getting a hand or foot caught and being dragged.

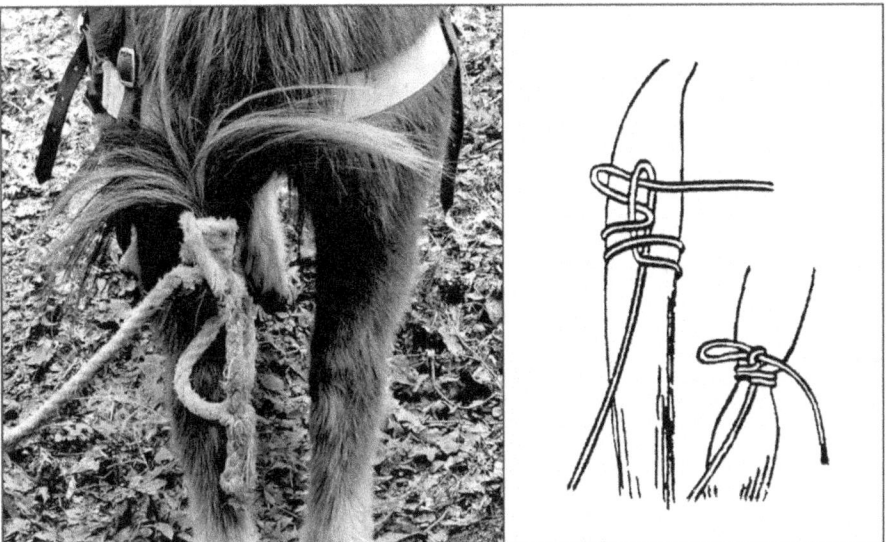

Figure 6-7. Tail Knot

- The rider should **never** move a pack string faster than the animals in it can comfortably navigate obstacles or difficult terrain. He should also keep in mind that the pack animals are carrying dead weight and, often, heavy loads. It is often preferable to give the string more slack when traversing an obstacle so the animals can pick their own way.
- It is preferable to have **two riders** per pack string—a puller and a drag man. The puller observes the trail to the front to anticipate problems. More important, the drag man observes the pack string and assists with any problems that may arise, such as a shifted load.
- If a rider is alone and does not have someone to watch the string for him, he can ride in a figure 8 to get a good look at the string.
- The rider should **always** be wary at halts. The majority of accidents occur at halts when the animals have the freedom to mill about and can become entangled. The rider should **never** allow a lead line to run under the tail of any animal, pack or ridden.
- If negotiating dense terrain (timber or rocks) and the pack animals choose a different route than the lead animal, it is better to **drop** the lead line and **recatch** the string than it is to become entangled in an obstacle with the string.
- The rider should **never** allow the pack string to get in front of him because if the string should become startled, the rider could get caught up in them.

- The normal distance between the lead pack animal and the rider's animal varies according to terrain and the animals' training and experience. A rule of thumb is the lead pack animal's **nose** should be even with the back of the rider's **flank** while traveling over easy terrain.

COMBAT CONSIDERATIONS

6-42. There are too many types of missions and units that could be assigned to mounted duty to allow this manual to encompass all aspects of military riding. Weapons, climatic conditions, and table of organization and equipment (TOE) will vary too greatly for this chapter to provide specific standing operating procedures (SOPs) and doctrine capable of covering all applications. This document will assist the user in formulating his procedures with general guidance.

WEAPONS

6-43. Individual weapons are as important to the mounted Soldier, regardless of his duty, as they are to the foot Soldier. They must be ready to be brought into action at all times. Achieving this state of readiness from horseback poses certain difficulties. A Soldier normally will carry his weapon in his hands when in a high state of readiness. While mounted, a Soldier's hands are often occupied. Also, there is the problem of keeping the weapon free of the animal and the equipment surrounding the rider. For a rider, there is also the ever-present danger of being separated from his animal. So a primary weapon cannot be kept attached to an animal.

Side Arms

6-44. There is a very real requirement for all mounted personnel to be issued a side arm. The preferred method of carry is the shoulder holster. The holster serves two functions: one—the weapon will be out of the way of tack and lines, always making it instantly accessible; two—the weapon will always be with the Soldier. Due to the large number of duties that mounted Soldiers must perform with their hands (to include tactical movement), a side arm, because of its ever-present availability, becomes a necessity.

Personal Weapons

6-45. Although the use of animals provides a commander a valuable asset for getting individuals and equipment to a battle, fighting from horseback is not considered a primary function of mounted Soldiers today. However, when a unit is in a hostile environment, it must be ready to fight at any time. For a mounted unit, this responsibility includes while on the move. As stated, side arms are a requirement for mounted Soldiers but they are insufficient as a primary weapon.

6-46. The standard weapons of the U.S. military (M16A2, M60, M240 MG, M203, and M249 squad automatic weapon [SAW]) have a serious defect in their size for mounted operations. It is difficult to handle the reins of a horse and a lead line while holding a large rifle. These weapons also demand a certain degree of accuracy that is next to impossible to achieve from horseback. To compromise between effective firepower and effective size, carbines are recommended. U.S. M16 variants such as the M4 are acceptable or, if operations were conducted in a UW environment, an AK folding stock variant would be acceptable. The ability of a submachine gun to lay a heavy base of fire from an unstable position makes it a valuable choice. A selective fire weapon with a folding stock, extended for accuracy when required, is the ideal choice. A very good choice of immediate suppression weapon for the mounted unit is the U.S. M79 grenade launcher. It is much shorter and lighter than the M203 and can be maneuvered with one hand. The launcher's compact size allows it to be placed in a scabbard and still be quickly brought to bear. Several M79s dispersed through a moving unit would greatly improve that unit's chance of surviving an ambush.

6-47. Another consideration would be the adoption of shotguns as standard weapons for mounted troops conducting operations in dense terrain. Their unequaled killing ability at close range and less severe accuracy requirement would make them a good choice as a weapon for mounted troops. The M249 SAW would be the best choice for a general fire support weapon because of its size. The box magazine of the M249 is recommended because a belt of ammunition is too unwieldy around animals. The main problem with any of the weapons described is how their size relates to how they can be carried effectively and still be brought into action when needed. Larger weapon systems 50 pounds or less, such as the M60, M240, and M249 MG, should be laid across the top of the rucksacks that are attached to the animals by using a packer's knot to secure, using 550 cord, or by using straps with quick release for easy access to weapon systems. As discussed before, the selection of carbine-style or folding-stock weapons goes a long way in helping with this problem. However, further mention must be made about how these weapons should be carried. As stated before, a rider can use a scabbard but only when enemy contact is highly unlikely, such as traveling in a secure area. A cross-chest carry with a top-mounted sling is the best choice. Great care must be made that weapons **do not** endanger the rider or his mount by becoming entangled in reins and lead lines or by hitting the animal.

ADDITIONAL WEAPON CONSIDERATIONS

6-48. The nature of modern battle dictates that mounted units carry more types of weaponry than the cavalry of old. Mortars, antiarmor weapons, air defense weapons, and sniper weapons are just a few that must be considered. Again, the METT-TC factors determine what weapons will be carried and in what manner. This manual discusses only a few typical weapons and generic considerations of each.

Antiarmor Weapons

6-49. U.S. antiarmor weaponry of any effect is generally too cumbersome to be carried on horseback. The possible exception is the M72 light antitank weapon (LAW). The rider can use the LAW in many different roles, but it is sufficient against main armor vehicles in only selected manners. He can conveniently attach one or two LAWs to the rear of a saddle, behind the seat, and rig them to be quickly released for action. Backblast must be considered if LAWs are fired around animals. The larger antiarmor weapons (M47 Dragon and tube-launched, optically tracked, wire-guided missile [TOW]) are, in all respects, too heavy and large to be mounted with a rider, but can be mounted on a pack animal. The rider must keep in mind, though, that these weapons cannot be brought quickly to bear if needed.

Air Defense Weapons

6-50. Air defense weapons suitable for mounted operations are the man-portable generation of weapons (for example, the Stinger or SA-7 GRAIL). These are too unwieldy to be carried for any distance on a ridden animal but can be packed easily on a pack animal with consideration given to speedy access. They should be placed within the pack animal strings where the qualified users are as well.

Sniper Weapons

6-51. Sniper weapons are not fired from a mounted position and generally not from a hasty position. They usually will not have to be brought quickly to bear. The sensitivity of weapon sighting mechanisms demands that they be protected when packed on animals.

PERSONAL EQUIPMENT

6-52. Standard U.S. personal equipment will serve a mounted Soldier well when certain considerations are given. In most cases, an assault vest is preferable to the web-gear style because it fits closer to the body. The Soldier can carry all items higher and out of the way of lines and reins. A standard all-purpose, lightweight, individual carrying equipment (ALICE) pack is too large and heavy for the mounted Soldier to carry. However, it adapts easily to being packed on an animal. The Soldier can place the ALICE pack on packsaddles just like panniers. He should carry all essential and sensitive items in his LCE. Using saddlebags and a small day-pack style rucksack can greatly enhance his ability to survive if his ALICE pack gets lost.

6-53. The Soldier will need to tailor his mission and personal gear to meet the weight limitations while using pack animals as a transport. A mounted Soldier carries the essential items with him. These include, but are not limited to, the following:

- Knife.
- Water.
- Rope.
- Necessary bag.
- Personal hygiene kit.
- Required first-aid supplies.
- Compass.
- Chemlights, all colors, to include infrared.

ADDITIONAL EQUIPMENT

6-54. Modern combat operations depend on communications. If a Soldier is designated to carry unit radio equipment, it must be carried with him. For this reason, commanders should make every effort to provide a mounted unit with the smallest and lightest communications equipment.

NOTE: Soldiers must never carry sensitive or classified communications items on a pack animal.

Chapter 7

Techniques and Procedures

It is easier to demonstrate how to pack an animal than it is to try to explain how to do it. The **only** way to learn this skill is by attending a school on packing or spending time with a knowledgeable person who can demonstrate how it is done.

TYING AND USING KNOTS

7-1. There are a variety of knots useful in packing; this section introduces a few and is by no means inclusive. There are several knots most frequently used and the Soldier may perhaps know, or learn, of others equally useful. As horses and packing become more familiar, the Soldier may even come up with some of his own inventions. There is no one perfect way to throw a hitch or tie a sling; the Soldier should use what feels comfortable and works best with the horse equipment and the load that has to be packed.

7-2. A Soldier cannot learn to tie a knot just by reading about it. As with packing and horse handling, the only way to learn this skill is by doing it. Table 7-1, pages 7-2 and 7-3, explains and illustrates several commonly used knots. The recommended method of learning is for the Soldier to get a length of rope and practice tying the various knots until he becomes proficient.

WRAPPING CARGO WITH A MANTEE

7-3. A mantee is a tarp that is either lashed over the top of the load secured on a packsaddle or used to wrap up cargo that is to be placed or slung on a packsaddle. Mantee loads are more versatile than pannier loads. The size and shape of a load can be adjusted to the items packed, and weights can be adjusted daily to each animal's capacity without wasting space. Mantees are easily loaded by one person and, although they must balance, their shape permits considerable adjustment on the animal without unloading or rearranging gear. Mantee equipment is simple. The packer needs only two sheets of canvas and two cargo ropes. Properly folded, a mantee holds a load together at least as well as a pannier and is often more weatherproof. Cargo could be duffel bags, kit bags, sacks of grain, or hay for the animals. Large duffel and sleeping bags that are to be slung on a pack load should be wrapped regardless of the type of material with which they are made. If the packer slings the bags next to a pack animal without wrapping them, they will pick up the sweat and oil off the animal. Besides soiling the articles, the bags will end up smelling like a sweaty packhorse.

Type of Knot	Knot Uses
Figure 4 Quick Release	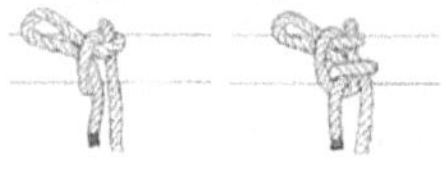 Used to tie a horse to the tree, high line, and hauling system for crossing bodies of water. No matter how hard the horse pulls, this knot will stay secure and still be easy to untie.
Half Hitch	 Has many uses but is mostly used during the lashing process around the mantee or to secure the end of the figure 4 knot.
Bowline	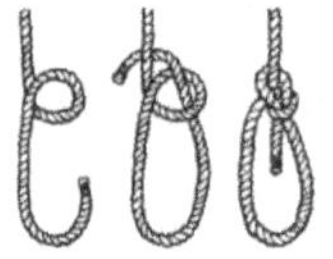 Used in various ways. One of the best knots for forming a single loop that will not tighten or slip under strain. This is the **only** knot that should ever be tied around an animal's neck. The bowline forms a loop that may be of any length desired. (When tying a horse to a tree or a picket pin, the Soldier should use the bowline since it will not tighten.) The loop remains loose and will not wind up if the animal walks in a circle around the tree or picket pin.
Clove Hitch	 Used to fasten a rope to a pole, bucks or arches on packsaddles, posts, or similar objects. It can be tied at the end of a rope or at any point along the length of a rope.
Timber Hitch	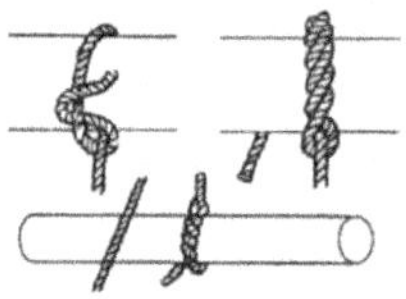 Used for moving heavy timbers or poles. The more tension applied, the tighter the hitch becomes. It will not slip, but will loosen easily when released.
Sheepshank	 Used to shorten a rope without cutting it. Also used to take the strain off a weak spot in a rope. It is a temporary knot unless the eyes are fastened to the standing part on each end of the knot.
Packers	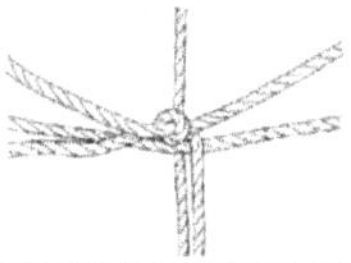 Used to secure the cargo rope around the mantee, barrel hitch, and basket hitch. It is secured with a half hitch.

Figure 7-1. Commonly Used Knots

Type of Knot		Knot Uses
Butterfly		Used to form a fixed loop or loops in place along the length of a rope without using the ends of the rope. It can be used to attach the middleman on a climbing party, tighten installed ropes, and make tie points on a picket line to attach the lead line of the pack animals.
Cat's Paw		Used to form a loop along the length of a rope without using the ends of the rope. It can be used to tighten installed ropes and to make tie points on a picket line.
Dutchman and Double Dutchman	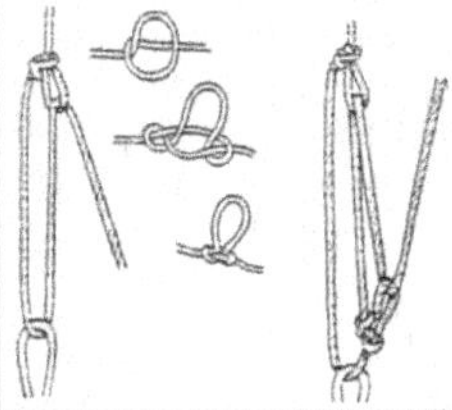	Used to take the place of a pulley. The Dutchman can be used to make tie points on a picket line or to secure the high line around a tree. It can also be used to secure the hauling system when crossing rivers.
Locking	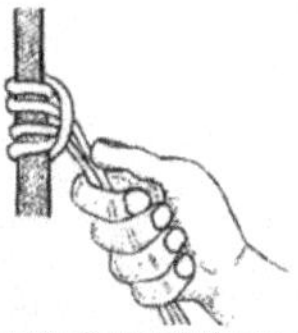	Used to secure billets on the offside of packsaddles.
Square		Used only in simple applications such as to tie packages and for binding rolls. It is easy to tie, will not jam, and is always easy to untie.

Figure 7-1. Commonly Used Knots (Continued)

MANTEE MATERIALS

7-4. Most packers use an 8- by 8-foot square of 12 to 18 ounces untreated canvas for their mantees. Lighter canvas tears and abrades too easily, and heavier canvas gets too stiff to fold when wet or frozen. The packer should not use tarps of synthetics. Plastics and nylon are too slippery to hold corners and tend to snag on brush. Mantee cargo ropes are made from waxed manila or soft-twisted, spun nylon rope with a 3/8-inch diameter. These ropes should be 36 feet long, back-spliced on one end and eye-spliced on the other end. The advantages to nylon are that it is stronger, lasts longer, is easier on the hands, and does not soak up water that can freeze. Rope stretch will not be a problem with the tension that is used. There will be more give in the load that is being packed than the rope. Also, spun nylon holds knots well, particularly after it has begun to fray.

MANTEE LIMITATIONS

7-5. If the packer does not properly tie a mantee, he can wind up trickling gear along the trail. However, tying the mantee can be easily learned and once mastered is easier and faster than loading panniers. Also, mantee gear is relatively inaccessible. To get something from a mantee, the packer must unload the horse or mule and take the mantee apart. Good planning can reduce this problem, but if the packer has a string of several horses or mules, it would be wise to pack one with panniers for ease of obtaining equipment used frequently.

HOW TO LOAD A MANTEE

7-6. If a packer can wrap a present, he can use a mantee on any item he wants to put on a pack animal. He must remember that the side next to the animal has to be flat and smooth. As in wrapping a present, the packer should take pride in the appearance of the finished package. The packer should drop two mantees so that the load will be balanced. The loads should weigh within five pounds of each other. The packer matches each item on one mantee with an item of equal weight on the other. He shapes the load to the practical limits of the pack animal and the design of the packsaddle. The packer loads the heavy item at a "third and a third." This means that the center of the heavy item should be one-third of the way down from the top of the load and should be in the center third of the load coming out from the pack animal. In this position, the weight rides on the tree of the saddle and directly under the sling rope when the packer uses a basket hitch. By focusing the weight at "a third and a third," the packer ensures that two loads of equal weight actually balance. For example, even though the weight is the same, if a load with the heavy weight at the bottom is loaded on the left side and a load of equal weight but with the weight concentrated at the top is loaded on the right side, the saddle will sag to the left. The gear should be arranged along the diagonal of each mantee ensuring no items have sharp edging that could damage the mantee or injure the animal.

7-7. When packing, the packer lays two mantees flat on the ground side by side. One will be for his onside and one for the offside. Cargo rope should be placed at the top of the mantee. The packer determines what his load will be and separates items into two separate piles of the same weight. The two piles should be within 3 to 5 pounds of each other. This method will give him a balanced load on the animal. The packer places one pile of the cargo on the onside mantee. He organizes (centers) the cargo with the heaviest weight to the bottom diagonally on the mantee. Figures 7-2 through 7-6, pages 7-5 through 7-10, illustrate how the packer uses four folds to mantee the cargo and also how to secure the mantee with the cargo rope.

FM 3-05.213

The packer stands near the bottom of the load. He folds the top corner of the mantee located closest to the bottom of the load, near the heaviest part of the load up to the center of the load.

He pulls the corner tight with tension toward the top of the load. He places his left knee on the top corner to hold tight.

Figure 7-2. First Fold of a Mantee

With his knee still on the mantee, the packer turns his body to the right side, reaches out and grabs the right corner.

He pulls up and over the first fold, keeping tension on corner of mantee to ensure it is stretched tight. He ensures any mantee hanging out on the corner is dressed up and tucked in so the corner has a squared look.

He brings the rest of the right side of the mantee up and over the cargo.

Figure 7-3. Second Fold of a Mantee

7-5

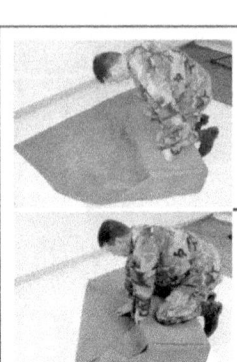 The packer rotates his body to the left side of the cargo, keeping his knees on top of second fold to keep tension on first and second folds.

 He reaches out and grabs the left corner and folds it in half.

 He brings the left corner up with tension, tucking in the left corner near the cargo.

 He brings the third (left) fold (remainder of fold) to the center of the cargo, keeping tension on the fold.

 He places his left knee in the center to hold the fold.

 He rotates his body placing his left foot on top of the load and maintaining pressure on load.

Figure 7-4. Third Fold of a Mantee

FM 3-05.213

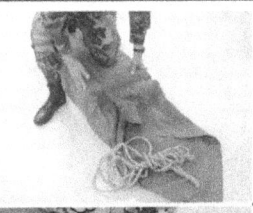

The packer continues to turn until he is facing the top of the load and places his left knee on the load to hold tension.

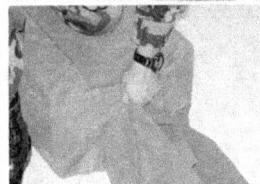

He reaches out and grabs the right and left corners, tucking and dressing the mantee.

He reaches out, grabs the bottom corner of the mantee and folds it toward him, ensuring the flap is as wide as the load to form a rain flap.

He tucks the bottom of the rain flap in so it ends up at half the distance down the cargo.

Figure 7-5. Fourth Fold of a Mantee

FM 3-05.213

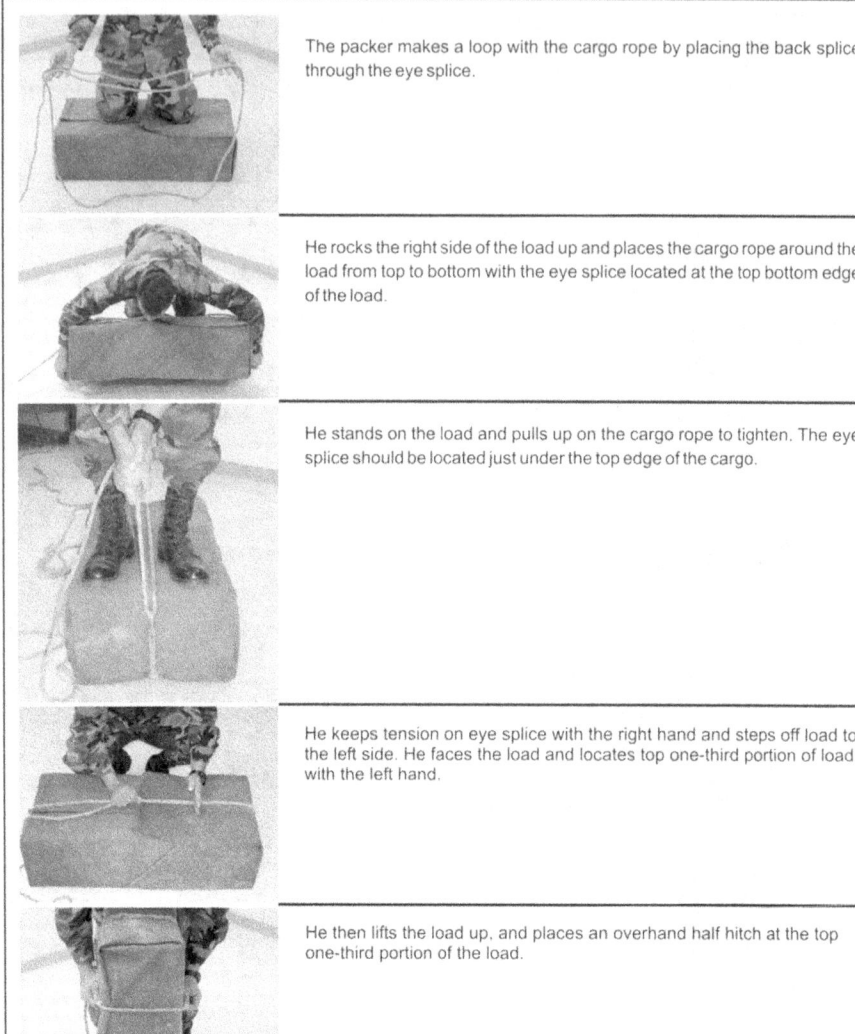

Figure 7-6. How to Secure a Mantee to Cargo With Rope

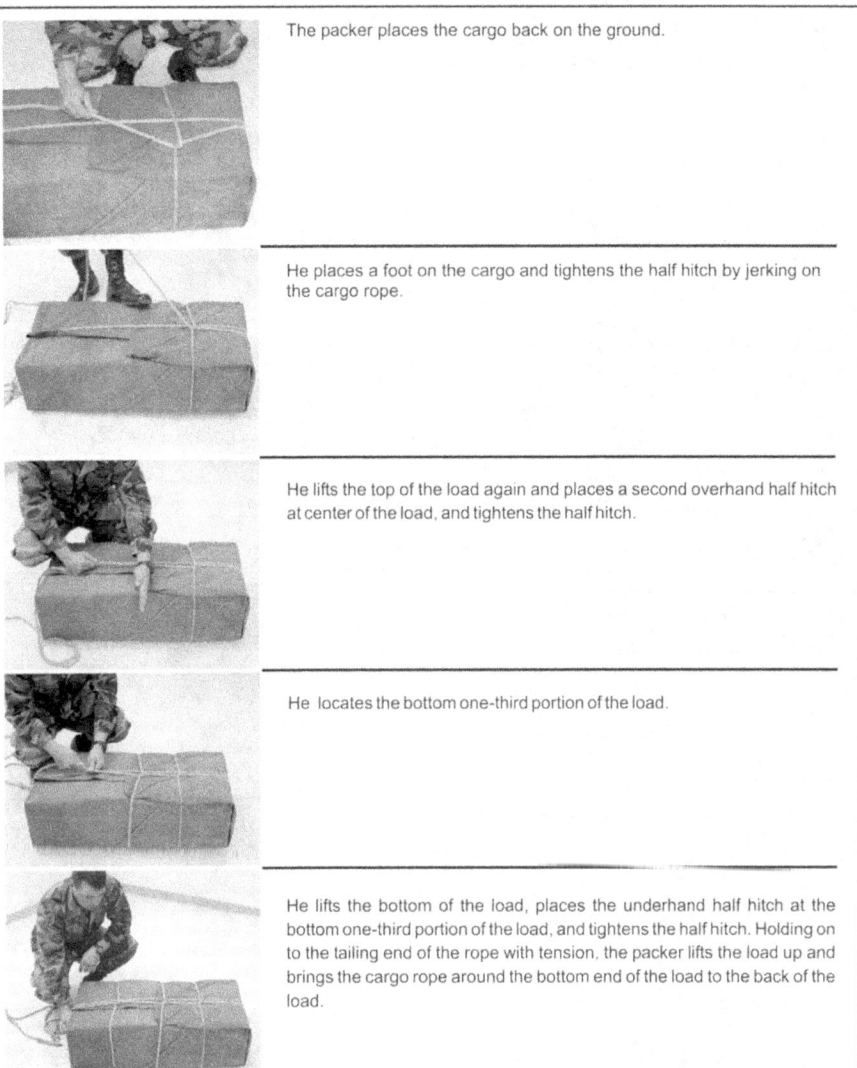

Figure 7-6. How to Secure a Mantee to Cargo With Rope (Continued)

FM 3-05.213

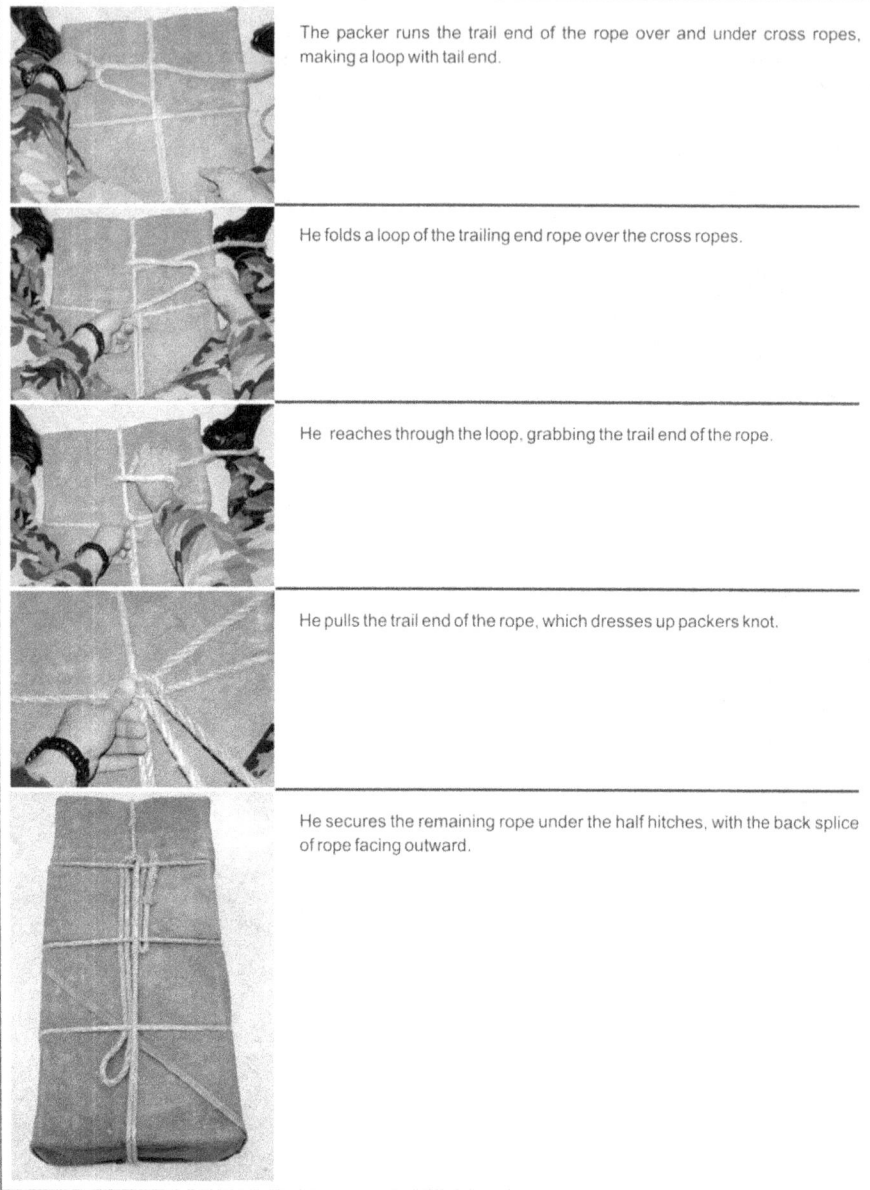

Figure 7-6. How to Secure a Mantee to Cargo With Rope (Continued)

NOTE: The packer then slings the load using a basket or barrel hitch. He can also use short ropes to loop the load to the bucks. He then ties off the load with a lash rope and box hitch.

BUILDING LOADS

7-8. Hard pack loads (Figure 7-7) are easier to mantee because of their shape. Hard pack loads consist of meal, ready to eat (MRE) boxes, ammunition boxes, water cans, and hard plastic cases, which all have a uniform shape.

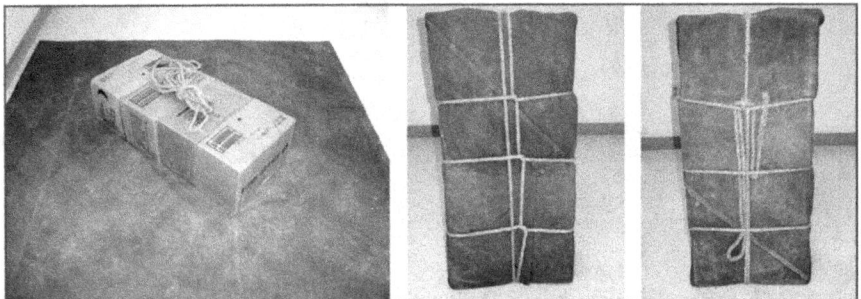

Figure 7-7. Hard Pack Load

7-9. The packer mantees soft pack loads the same as hard pack loads. The soft pack load will be harder to organize and bulkier with no uniformity in shapes of different sizes. Soft pack items consist of kit bags, medical bags, rucksacks, demo bags, duffle bags, and a few hard items such as an MRE case or water can (Figure 7-8).

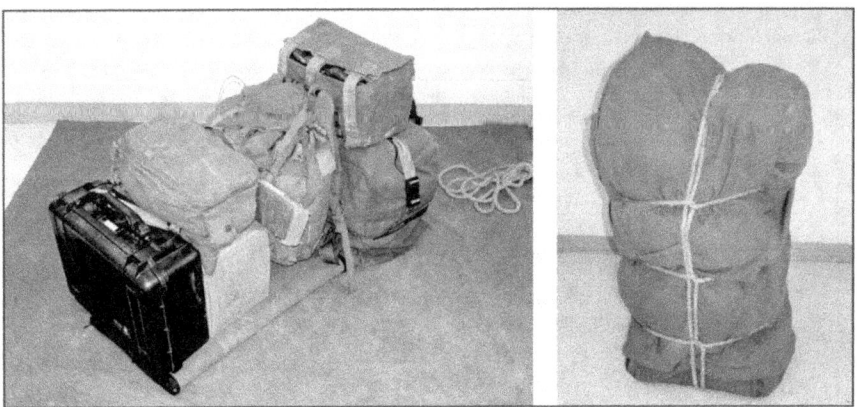

Figure 7-8. Soft Pack Load

SPECIAL WEAPONS LOADS

7-10. Weapons systems, such as the M2, MK-19, 60-millimeter (mm) and 81-mm mortar, can either be put in a mantee or into panniers designed for a heavy weapon system. If the packer uses a mantee, he must ensure all metal parts have padding to reduce the noise from movement. He ensures that the smooth side of the weapon is on the bottom of the mantee where it will be placed on the pack animal's back. He also equally divides the weapon system's weight when using the mantee. Table 7-1 shows one way to equally divide the weapon for the onside and offside of the animal. Ammunition may need to be loaded on different animals or broken down to equal the weight and packed with the weapon system to ensure loads are balanced on both sides of the animal. The packer also needs to throw five half hitches around the manteed weapon system for more security, instead of three as is done for the cargo loads.

Table 7-1. Onside and Offside Loads

Weapon	Onside of Load	Offside of Load
M2 Cal .50 Machine Gun	M2 Without Barrel, 61 pounds Cradle, 21 pounds T&E, 2.5 pounds Gloves, 1 pound Headspace and Timing Key, 2 ounces **Total Weight = 85.7 Pounds** NOTE: 1 each, case of cartridges, cal .50 Ball M33 linked, 77 pounds.	Tripod, 44 pounds Barrel, 24 pounds Barrel, 24 pounds **Total Weight = 92 Pounds**
MK-19 40-mm Grenade Machine Gun (MOD 3)	MK-19, 75.6 pounds T&E, 2.5 pounds **Total Weight = 78.1 Pounds** NOTE: 48 rounds of M548 in metal container, 62 pounds.	Tripod, 41.5 pounds Cradle, 21 pounds 1/3 Ammunition (Ammo) box, 16 rounds, 19 pounds **Total Weight = 81.5 Pounds**
M224/225, 60-mm Mortar	Mortar, 18 pounds M64A1 Sight, 2.5 pounds Bore Brush, 2.5 pounds M2/M2A2 Aiming Circle With Accessories, 12 pounds Without Batteries, 9 pounds **Total Weight = 32 Pounds** NOTE: High-explosive (HE) M888 ammo, 3.41 pounds per round. For hand-held mode, M8 base plate weighs 3.6 pounds.	Bipod, 15.2 pounds Base Plate, 14.4 pounds M14 Aiming Post With Case, 6 pounds **Total Weight = 35.6 Pounds**
M252, 81-mm Medium Extended Range Mortar	Mortar Barrel, 35 pounds Mortar Mount Assembly, 27 pounds **Total Weight = 62 Pounds** NOTE: HE M821 ammo, 9.5 pounds per round.	Base Plate, 29.5 pounds M64 Sight, 2.5 pounds Bore Brush, 3 pounds M2/M2A2 Aiming Circle With Accessories, 12 pounds Without Batteries, 9 pounds M14 Aiming Post With Case, 6 pounds **Total Weight = 50 Pounds**

NOTE: Ammunition or other equipment may be needed to adjust weight to the offside when packing.

7-11. Weapon systems like the AT-4s, LAWs, or Javelins (weights are shown in the special segment box below) can either be strapped together and left exposed or put in a mantee as in Figure 7-9. If the packer decides to mantee the weapon system, he must ensure all metal parts have padding to reduce the noise. He should also make sure the smooth side of the weapon is on the bottom of the mantee that will be placed on the pack animal's flanks. The packer makes sure to divide the weapon system's weight equally for the onside and offside. For more security, he should use five half hitches (three overhand and two underhand) around the mantee instead of the typical three.

Figure 7-9. Exposed Javelins and AT-4s in a Mantee

M98A1 Javelin, Surface Attack Guided Missile, 49 pounds
M72A2 Light Antitank Weapon (LAW), 5.1 pounds
M72A3 Light Antitank Weapon (LAW), 5.5 pounds
M136 Antiarmor Weapon (AT-4), 14.8 pounds

7-12. Top-loaded weapon systems are lighter in weight. The packer can secure the M249, M60, M240, and AT-4s across the top of two rucks or cargo packs that are in a mantee and hung on the onside and offside of the animal (Figure 7-10, page 7-14). These weapon systems will be secured with quick releases or straps to the mantee or rucksacks.

NOTE: No top-loaded weapon system should weigh over 50 pounds.

7-13. Rucksacks can easily be attached to the Decker or sawbuck saddle by using the single-point release system used for airborne operations. The packer attaches the single-point release system to the rucksack in the same configuration as if he were going to jump. He takes the two snap hooks and attaches to the saddle on the arches of the Decker saddle (Figure 7-11, page 7-14) or by running 550 cord around the saddle crosspieces on the sawbuck saddle.

Figure 7-10. Top-Loading a Weapon System

Figure 7-11. Rucksacks Attached to Decker Saddle

7-14. Planning is the most important part of packing. Whether a person is packing one animal or twenty, it takes a plan of attack. In reality, a pack animal is packed in the packer's mind before the actual pack is ever loaded on the animal. A packer should decide which animal will carry which load best, which will be the lead pack, and which can be trusted to stand without trying to lie down or roll while the others are loaded.

7-15. A very good method to follow in planning the pack loads is to lay out several tarps or mantees on the ground to keep the items clean and dry. If the packer has several personnel whose gear is being loaded in the packs, he can have them put it all on the tarps. The packer should explain that once he starts to lay out the packs, they cannot take anything out or add to the pile of gear without asking him first. Any item taken out of a duffel bag that the packer has already hefted for weight will throw off the weight of the pack. This amount could be just enough weight variation to either cause sores on the pack animal or slip the pack on the trail.

7-16. When preparing to pack, the packer loads each individual's equipment onto one animal as neatly as possible, and if more than one animal is needed, he keeps them together in the string. This practice makes things simple at overnight halts and when the final destination is reached, so one individual does not have to go through the panniers of several different animals searching for his equipment.

7-17. One of the many secrets of packing—possibly the most important—is to keep the weight down at the bottom of the pack. This gives the packer a good solid foundation on which to build the rest of the load. A good load for most horses or mules (1,100- to 1,200-pound packhorses and 800- to 1,000-pound mules) is 160 to 170 pounds. However, if there are some small animals in the pack string, this amount could be too much for them. In extreme circumstances, the packer can load an animal with up to 250 pounds; however, this amount would limit his speed and endurance. The horse gear (shoes, nails, hobbles, bells, picket chains, ropes) usually will fit in one set of panniers (two panniers). The packer must remember this set of panniers well, as it will be the first item needed when he unpacks at the destination. He should try to keep all the camp tools (shovel, ax, saw, currycomb, brush) on this pack load; it will save a lot of time when he arrives at the camp for the night. The packer will know exactly where all his tools are located.

7-18. When loading the bottom of box panniers, the packer wants everything to fit as snugly as possible to keep from rattling. A pile of gunnysacks close at hand is good for chinking up any rattles. The packer can distribute the items through several sets of panniers, if necessary. Two layers of heavy items in a pannier are a big bottom load. If the packer has plenty of pack animals and box panniers, one layer of heavy items in several boxes will be better than trying to get it all on one pannier.

7-19. Canvas panniers come in all sizes, shapes, and forms. Their general use in packing is to load them with bulky items that will not fit the box panniers. The loading method as mentioned before is the same: the weight should stay down at the bottom. Just because a canvas pannier sometimes looks like a big sack, the packer should not make the serious mistake of loading it like a grocery bag. There is nothing between the load in the canvas pannier and the pack animal but a thin layer of canvas. The packer loads the canvas pannier so it is smooth and flat on the side next to the pack animal. If there are a lot of odd-shaped sleeping bags and duffel bags, here is a good place to put them. This method is also the best way to pack the grain for the animals. Having numbers on all the panniers is a great help in keeping them in sets of two. Whether there are numbers on the panniers or not, the packer should keep the sets together when he has finished loading.

7-20. When the packer has all the supplies, horse gear, grain, and other things packed in the panniers and any sling loads wrapped, he is ready to finish the pack loads with personal gear of the party. The packer selects two items of equal weight and puts one in each pannier, all the time keeping in mind what each pannier weighs. Next, he may need a top pack. He selects two duffel bags or sleeping bags of equal weight and places these in each pannier. Now is a good time to recheck the weight of his panniers to make sure they are still equal. The packer should have a scale that he can suspend from a tree limb to weigh the panniers; a difference of only a few pounds between them can make all the difference once the animal starts down the trail. If the packer feels that both panniers together weigh under 150 pounds, he should find a third sleeping bag or duffel bag for a center load. He then lays this third item across both panniers, showing it as the center load. He places a mantee and lash rope over the entire pile and allows no one to disturb it until he is ready to load it on a pack animal.

7-21. After the packer has gone through the pile of things to be packed, there will probably be several items that look like they will not fit any of the loads that have been laid out. If they are small and light enough to make another pack load, the packer can often find a place for them as he packs out the other loads. He will have to allow space for these items while loading the other panniers or he will end up with one or two things left out.

SADDLING

7-22. This section explains techniques and lessons learned that will keep a rider in good stead. (Chapter 5 provides saddling details.) There are quite a few different types of packsaddles, and each has to be rigged differently to get the pack on it. These saddles may have any number of different riggings and may or may not have quarter straps. Whatever the type, all saddles must properly fit the pack animal.

7-23. The padding between the pack animal and the saddle is a vital piece of pack gear. Pack pads as a rule are larger and thicker than regular riding pads. Once the animal is packed and the load rests on these pads, they will stay there until the animal is unpacked. Pack pads are made of many materials, the best being one that will stay soft and not pack down. The most commonly used pads have fiber, fleece, felt, or hair sewn to a light canvas.

7-24. Whatever type of pad is used, it should be checked thoroughly for any foreign objects before placing it on the animal. Sweat will build up but can be scratched off with the currycomb. Wet pads will gall or cause sores on a pack animal. The packer lays the saddle pads out where they can air out and dry when possible. He should always try to place the pack pads square on the packhorse, leaving at least 4 inches to the front of the packsaddle. Depending on the shape of the pack animal's back, the packer may need more than one pack pad. The best way to gauge is to set the packsaddle on the animal, then check the clearance between the saddle and the animal's withers; there should be enough space for two fingers.

7-25. The packer allows for the saddle settling down on the pads after the animal is loaded. If there is any chance of the saddle forks coming in contact with the top of the withers, he puts another pack pad or cheater pad under the saddle. He also makes sure not to put too much padding on the withers and cause the packsaddle to pinch from being too thick. Most good pack pads are thin down the center of the pads. The aim in a high-withered animal is to raise the packsaddle off the tops of the withers. Some packers use a cheater pad as explained in Chapter 5, page 5-8.

7-26. Now is the time to again check the pack animals to see if the saddles on them are rigged for the type load the animals will be carrying. If the packer has one or two sling loads, he knows to have that many sling ropes on the saddles and have them tied off ready to use in saddling the pack animal. He rechecks the pack pads to make sure they are loose over the withers of the animal and that the packsaddle sits in the middle of the pads.

7-27. During all the loading of the panniers and other gear, the packer should **not** lead the pack animal around to pick up parts of the load. It is a very serious mistake. He should have all the gear at hand that will be loaded onto the animal in one central location where he will be doing the loading. Once the packer begins loading an animal, he waits until the pack load is lashed down before he moves the animal, or he will be picking up parts of the load from the ground after they have slipped off.

7-28. If the packer is going to move the pack animal from where he was saddled to where the pack load is sitting, he pulls the cinches up snug before he leads him to the pack load, then finishes tightening the cinches just before loading. This technique is called "untracking" the animal. If pack animals that have been saddled and standing with a loose cinch have the cinches tightened and are then loaded without moving around first, they could experience some discomfort from being pinched or having a fold in their skin caught under the cinch. This pain will often cause them to throw the whole load off. When the packer pulls the cinch up on a pack animal, he untracks the animal before he puts the pack load on him. Moving around a bit allows the cinch to settle into place, and the animal will be more comfortable. This practice also applies when the packing is completed. The packer should continuously check the animals and monitor how the loads are riding. He should especially watch the animal as it takes the first few steps after being packed; if the load looks like it is shifting, he makes whatever adjustments are necessary.

7-29. Whenever possible, the packer should get help in loading the panniers onto the animals, because when packing alone, he will have to keep moving from side to side. While the packer is moving to the opposite side, the load on one side is pulling the packsaddle over and pinching the animal's withers. With a heavy load, this could cause the animal enough discomfort to buck off the load or at least turn the packsaddle.

7-30. Some panniers have ropes or straps that the packer can adjust to set the height at which the panniers will sit on the pack animal; others are slung with the sling rope. Regardless of the method used, the panniers must be even in height on the pack animal. A pannier low on one side pulls the saddle in that direction and either causes a sore on the animal or slips the entire pack. After loading the panniers on the animal, the packer steps to the rear and checks that they are evenly placed.

7-31. With any center load, such as a duffel bag, the packer must make sure the opening is to the front of the animal. The motion of the animal usually shifts things around. With the opening to the front, the packer can see if any items are working their way out of the bag while on the trail. Regardless of what is center-packed, the packer should make sure that after the load is lashed down it will not rub the animal.

7-32. The packer pulls the mantee down evenly on both sides of the boxes and tucks in the ends around the panniers. He keeps checking to make sure the pigtail on the packsaddle is clear and out where he can tie into it. A good solid pack will look sloppy if the packer does not take the time to use the mantee neatly and tie it up right.

SLINGS AND HITCHES

7-33. The packer uses the sling to initially attach the load (panniers, mantee, hay) to the saddle. Once attached, he throws a mantee across it and uses a hitch to secure the entire load to the animal. He then ties the sling to the saddle and the hitch to the animal.

7-34. The basic purpose of a hitch is to secure the entire load to the packsaddle and the animal as a balanced unit and still not have to use 100 feet of rope. The packer then throws these hitches so they can be taken off the load with little effort.

7-35. The different means of tying a load down on a pack animal are not as confusing as they might sound. Terms such as a one-man diamond, half diamond, full diamond, double diamond, squaw hitch, box hitch, and many other means of tying are frequently used during packing. The packer does not have to learn them all, but if he learns one or two of the most commonly used hitches, he can tie down almost any load encountered.

7-36. To throw a hitch means just that. The packer throws the loops and coils of his hitch on the pack in such a way that when he gives the hitch its last pull, all the ropes pull tight. When he releases the hitch to unpack the animal, the packer does not have to spend time unwinding or untying knots in the lash rope.

7-37. Most new lash ropes are approximately 36 feet long. They come in many diameters and are made of several materials. In most cases, lash ropes are approximately 1/2 inch in diameter and either of manila or nylon material. Polyester rope makes the best lash ropes because wet weather makes grass or hemp ropes very difficult to handle.

7-38. If a packer uses wet slings or lash ropes made of hemp to pack an animal, he may not go very far up the trail before the slings and hitches start to dry and stretch out. This stretching will make his pack loose or uneven and could cause a considerable delay in reaching his destination. In camp, or wherever the mission leads, the packer should keep all hemp lash and sling ropes dry if at all possible. The ropes should not be thrown on the wet ground while packing an animal. They should be hung on a tree limb or laid on a tarp or dry place until ready for use.

7-39. The lash cinch (on the end of the lash rope) needs checking often for signs of fatigue in the cinch materials. It has an eye splice in one end and a crown splice (back splice) in the other end to simplify the makings of a hitch. The lash cinch is really, in most hitches, the beginning and the end of a hitch. The packer throws the lash cinch across a loaded pack animal, and very often, the person on the offside forgets to duck or watch for it to come over. It does not take many knots on the head to learn respect for this piece of pack gear.

7-40. Whatever hitch the packer uses on the pack load must hold the ends, sides, and bottom of the panniers, besides holding the top pack. If the packer balances the pack on the animal properly, the lash rope can hold the pack down and together. He cannot balance a pack with the lash rope. After the packer puts everything inside so it will not rattle or break and the wrapping is on, he is ready to lash. He wants something that looks good yet holds the wrapping and box. Handling a lash rope while throwing a hitch can be dangerous if the packer becomes tangled in the rope and the animal gets spooked or starts bucking. The packer should always make sure there are no coils around his feet or arms, as they could cause him to get dragged or seriously injured. The safest way to handle this situation is to keep the tail, or excess, rope thrown out to the right while throwing the hitch.

7-41. The packer should always check the pigtail on the packsaddle to make sure it is still clear and ready for use. Again, he checks the entire load, front and rear, by conducting a press check to ensure packs and saddles are in line with the animal and are riding balanced on the animal. If it looks like the entire pack load has shifted to one side, the packer lifts up on the low sides and sees if it will rock back straight. If it does not, he completely repacks the entire load. This rechecking will save the packer time on the trail repacking where he may not have help or a place to tie up. It can also prevent sores from developing on the animal. He checks any loose rope or mantee ends sticking out and tucks them in the wrap.

7-42. The methods of securing cargo are slings, used to attach the load to the saddle, and hitches, used to secure the entire load to the horse. If the packer ties the sling on a sawbuck saddle, he can tie the same sling onto a Decker saddle by running the ropes through the loops (arches) at the top of the saddle rather than behind the bucks. With a little imagination and ingenuity, the packer can adapt any of the slings and hitches to whatever type saddle (to include riding saddles) and load he is packing.

LOAD BALANCING TIPS

7-43. The packer should make sure the animal is standing on level ground before he begins loading. He should beware of an animal resting a hind foot. The packer should quickly get the second mantee or pannier on the horse or mule. This can be accomplished with two packers better than one. The packers should balance each pair of loads **after** it is on the animal. This can be done by rocking the packs vigorously in place, releasing them, and seeing that the D-rings center perfectly on the animal's pack. If loads are not balanced, the packer should, whenever possible, lower the light side rather than raising the heavy side. He can snug the load up to within an inch of the D-rings, regardless of how high or low he positions a load.

BARREL HITCH

7-44. If the saddle does not already have a sling rope attached to it, the packer ties one onto the sawbuck saddle by using a clove hitch in the center of the cargo rope. He ties onto the Decker using the eye splice and back splice of the cargo rope attached to the front arch with one rope and to the rear arch with another cargo rope. He ensures the sling rope is around 36 feet long, though this may vary according to the size of the cargo.

7-45. Starting on the onside of the animal, the packer makes a large loop at the forward end of the saddle, passes the rope behind the bucks to the rear, then makes another large loop at the rear of the saddle. He then brings the rope back behind the rear buck and down to the sling ring (Figure 7-12A).

NOTE: The packer repeats this procedure on the offside of the animal.

7-46. The packer slips the cargo through the loops of the sling rope, then tightens up by running the end around the section of rope behind the bucks, pulls tight, and then ties off onto the sling ring (Figure 7-12B).

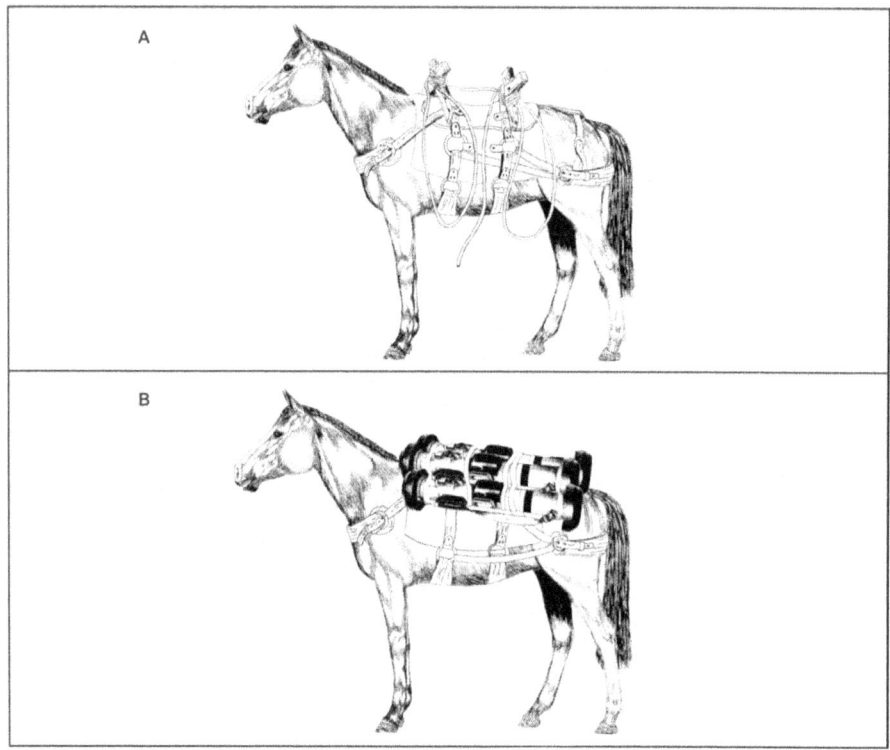

Figure 7-12. Barrel Hitch

FM 3-05.213

> **CAUTION**
> After running the ropes under the forks of a sawbuck saddle, the packer checks the clearance between the ropes and the withers to make sure the ropes will not rub the animal's back raw.

BASKET HITCH

7-47. As with the barrel hitch, the packer starts by tying a sling rope onto the saddle. He draws the rope around the cargo, through the back hoop, down between the cargo and the animal, then back out and up the center of the outside of the load.

7-48. The packer pulls one loop through the horizontal portion of the rope, then pulls another loop through that one. He pulls the first loop tight, then expands the second to fit around the bottom of the cargo. He allows enough slack so the bottom of the loop can pass through the cinch ring. He threads the end of the rope through this loop. The packer pulls the sling tight, then pulls another loop through the horizontal section and secures it with two half hitches (Figure 7-13).

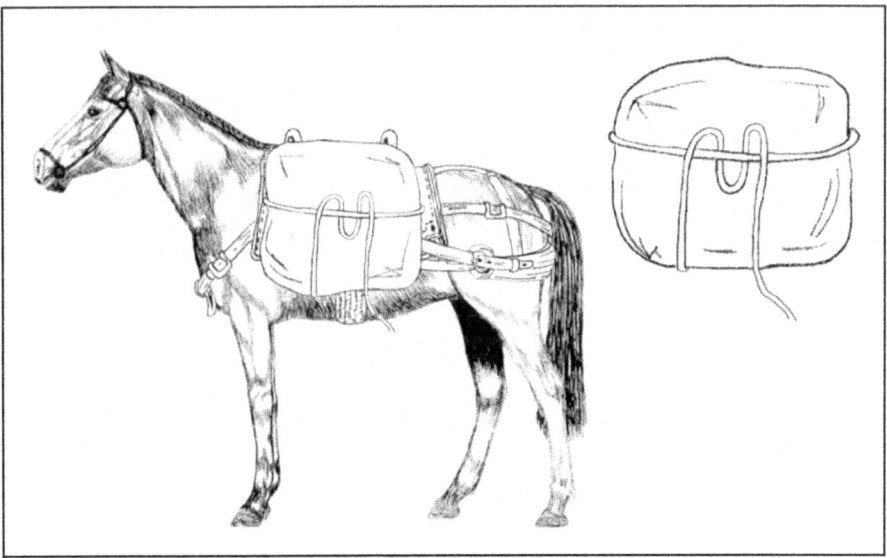

Figure 7-13. Basket Hitch

DIAMOND HITCH

7-49. The diamond is useful for soft loads. The one-man diamond is easiest to tie (Figure 7-14). Again, this hitch does not tie off anywhere on the saddle.

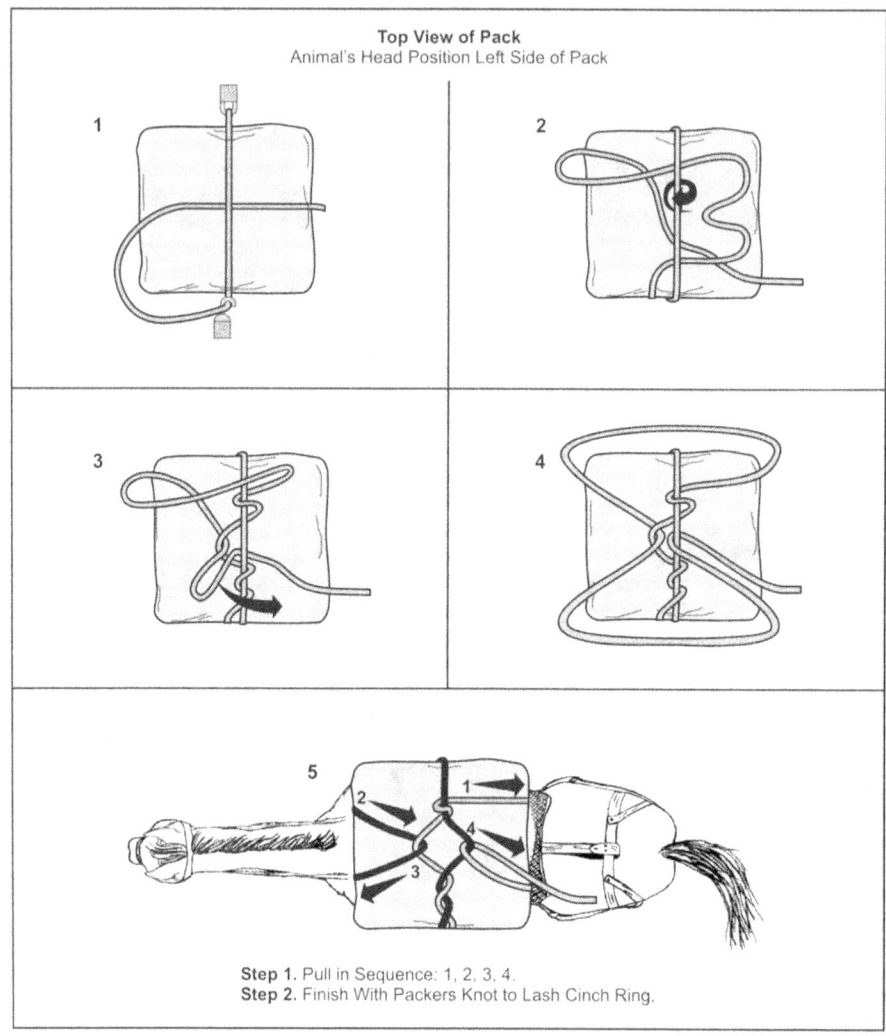

Figure 7-14. Diamond Hitch

7-50. This section is not meant to totally encompass slings and hitches—just enough of the basics to suffice for the most commonly encountered situations.

As stated before, a person cannot learn this skill by reading about it; he must find someone knowledgeable on the subject and have him demonstrate how it is done.

THE PACK STRING

7-51. Once the packer has first-class, well-balanced packs on the pack animals, he needs to make up the pack string. Being knowledgeable of how all his animals perform will help the packer greatly when he is selecting the lead animal. If the packer selects an animal that will not lead up on a slack rope, he will spend his time dragging this animal along all day. Many pack outfits using mules have a horse they call the Bell Mare. The outfit can put a bell on this particular horse and lead just the one animal and the rest will usually follow. Using horses as pack animals is another situation. Horses and mules form habits, just as teams do, as to what position in the string they should be. If the packer knows this position when tying a string together, he will have a well-organized pack string going up the trail. Often, one of the pack animals will have a very fragile load that needs special attention on the trail. The packer may find that this is the animal he wants to lead and should focus on the pack items when loading out the packs.

7-52. Personnel should always stay in the clear whenever they are afoot around pack animals tied together. Many things can happen, even with the most gentle pack stock. Two animals with box panniers can definitely put the squeeze to a person caught between them. It does not have to be the pack animal that a person is working on that causes the trouble. It can be any one of the animals in the pack string. If one jumps or bucks, the whole string, being tied together, has to go along. If a person is in the middle of several loaded pack animals, he should try to keep them standing apart and in a line. A person should never get in the middle of the pack string when they are bunched up.

7-53. A person should never get off his saddle horse into a crowd of spooked and bucking pack animals. He should try to keep them circling or headed straight out until they settle down. He does not stand a chance afoot until he has them where he can get a hold on the head of whichever animal is causing the problem. A person may feel helpless sitting on his saddle horse, watching all that gear being thrown off and trampled, but he must remember that he could very easily be the one that is getting trampled or kicked.

7-54. Tying the pack string together takes just a little horse sense. The packer wants to give the animals all the room possible between them and yet not have so much lead rope that they can get a leg over the rope when their heads are down. A good gauge for the length of rope is to use the hoof of the pack animal that the packer is tying off to. If the animal being tied off is standing roughly nose-to-rump to the other animal, and the lead rope is down about to the hoof on the lead pack animal, this length is about right.

7-55. The packer can tie the pigtail on a packsaddle in several ways. One opinion is to run it through the rear forks of the packsaddle and down to the front rigging rings. Installing a pigtail this way pulls only on the cinch and not on the packsaddle. A nylon or polyester piece of 3/8-inch rope braided at the rings makes a very substantial pigtail.

7-56. Oftentimes, it is a good idea to have a "breakaway" in the end of the pigtail where the packer will be tying the lead ropes of the pack string. The packer can make the breakaway with light baling twine (40 to 60 pounds) or 1/4-inch 80-pound cotton webbing used for airborne operations. Should there be an emergency where the pack animal needs to break loose, it can usually break this light twine.

7-57. When leading a string of pack animals, the packer should be all eyes. The first mile on the trail is the most important. This short distance helps the packer find out if the packs are tight and well-balanced and if he missed some rattles in the panniers. He will want to look back over the pack string about as much as he looks ahead. When the packer comes to a turn in the trail, he should get a side view of the packs. Often, he can see something coming loose that can easily be fixed, but if left unattended could cause a pack to slip or something to be lost on the trail.

7-58. One of the most common mistakes a green packer makes is not giving the pack string time. When the packer crosses a ditch, rock, or downed trees in the trail, his saddle horse most likely will just step over or around it and keep right on going. Possibly, the next pack animal (the lead) will keep up easily enough, but the second pack animal will have to really hurry to get across the obstacle. When it comes to the third animal, it will either have to jump or pull back and so on down the line. When the packer crosses or goes around any obstacle on the trail, he should slow up until all his pack animals have gotten around or over it.

7-59. If the pack string has steep country to climb, the lead rider should give the animals a lot of time to stop and get their breath. He should pick a good spot where all the animals are standing square on the trail. The rider cannot use his saddle horse as a gauge for when to stop to breathe the pack string. The saddle horse may be carrying a 200-pound load, but that load is in balance with him while he is climbing. The pack animal has dead weight and needs more air to pack his load up the hill.

7-60. On long trips and especially when climbing, periodic rest stops will be necessary, but the packer should not overdo them. Stopping for long periods on the trail can cause more packs to needlessly slip. If a stop for lunch or whatever is necessary, the pack string should be taken apart and each animal tied to a tree. If at all possible, the pack string should be kept in motion. The pack load is continuous on the animals and the quicker it can be taken off, the better for the animals.

7-61. When crossing a stream or any fresh water, the animals should be allowed to drink. The lead horse should not just ride in and stop, the rider should, if possible, make sure there is room for all the animals to drink at once. When the animals are done drinking, the lead rider should make sure none of them has a foot over the lead ropes as he starts out. Whenever the lead rider stops on the trail for even a short time, he should always be sure the pack animals are ready to go before moving out. Often, one animal will be spread out relieving himself and is in no position to step out.

7-62. One of the most irritating things in leading a pack string is to have a pack animal that is always going around a tree the opposite way the rest of the string is going. Sometimes, shortening the lead rope will correct this problem. If that does not work, the rider may have to lead him. This behavior is usually a sign of a green pack animal and quite often disappears as the animal learns what is expected of it. Others just do not ever seem to learn. If the pack string has one of these problem learners, the packer should consider leaving it behind on the next mission.

7-63. As the packer becomes more proficient in packing and leading the string, he will learn specific techniques that worked well and will apply lessons learned through past mistakes. The more common tips to remember when positioning animals in a string are as follows:

- Make sure the lead animal is one who has a proven ability to lead easily and will not kick at the animal behind him.
- Do not put the fastest horse or mule at the head of the string. Put a steady, slower animal in this position—one with a walking pace that the entire string will find comfortable, and who will not be running up on the lead saddle horse or trying to pass him.
- Place the more quicker, more agile animals toward the end of the string to take advantage of their ability to negotiate obstacles and keep the string closed up.
- Put progressively taller animals toward the end of the string to make it easier to see how rear loads are riding.
- Put a problem animal, like a puller who knows how to break pigtails or a round-back mule whose saddle slips, in the number two position, so the packer will not have to squeeze past the whole string to adjust it.
- Keep the strings short. Leading two or three animals is not difficult. Leading six to ten can be a real headache. Short strings of five or less move faster than long strings.
- In difficult terrain, spend more time looking back at the string and between each animal's ears to see if the D-rings of the cross bucks of the packsaddles are centered on every animal's back.
- Resolve little problems, like unbalanced loads, quickly with the pack animal in place in the string. For major adjustments, remove the animal from the string to fix the load.

CAMPSITES

7-64. If there are several animals in the outfit, and the handler does not picket all of them, he might find it almost impossible to keep the loose animals in the area if there is insufficient grass. Unless he spends the entire night watching them, the animals might wander off in search of good pasture and be gone in the morning. Grass and water go together. An animal without water will act much the same as one with insufficient grass. For this reason, every handler should sleep with his weapon and bridle. A spring of good, cold water is by far the best, but these springs should run water far enough **below** the camp for the stock. Another problem in most all pack or saddle stock is the tendency to want to return to the last camp, or worse yet, go home. This urge is more evident in them if the string should be heading toward home after a few days out on a mission.

SETTING UP CAMP

7-65. If the packer remembers which items are in which pack, he will know which load goes where in the camp area. It will save a lot of time and labor to be able to lead each pack animal to wherever its load goes. As soon as the pack is off the animal, the handler should loosen the cinches on the packsaddle.

7-66. Keeping the saddles and pads off the ground even in dry weather is a sign of good organization. It only takes a minute to lash a pole between two trees to put the saddles and pads on. A tree limb makes a good hanger for lash ropes and halters. When unsaddling an animal, the pad and saddle should stay together. If the saddle is marked for which horse it fits, the handler will not have to refit the pad the next time he saddles up. If the weather is clear, the handler should let the pads air out and dry before covering them with the mantees off the pack loads that evening. While unsaddling the animals, the handler should ruffle the hair on the backs of the animals and check for any tenderness or sores. If a sore-backed animal is noted, it should be taken care of before being turned out.

7-67. When setting up for the night, many pack gear items will serve a dual purpose. Box panniers turn into tables and stools, lash ropes turn into high lines and corrals, and mantees turn into ground tarps and covers.

THE HIGH LINE

7-68. The high line is a section of rope strung between two or more trees, with tie points for the horses (Figure 7-15, page 7-27). The knot used for these tie points (sometimes called the high-line knot) can be the pulley knot used for a single or double dutchman, a butterfly knot, or a cat's paw. Another more permanent and faster way than tying knots is to insert snap links woven into the high line before the mission at 8- to 10-foot intervals used as the tie points for the animals.

7-69. The high line can often be temporarily put up when first arriving in camp. It will serve as an ideal place to tie up the pack string while laying out the camp area. Lariats will also serve for the first high line until the handler can get a lash rope off the packs. Tying up to a line upon arriving at camp can often prevent damaging equipment and gear on a pack. Tying keeps the pack animal from getting under low limbs or trying to rub the pack off against the tree he is tied to. A very frequent mistake is tying an animal to a small dead tree. Quite often just a light pull on the dead tree in the right direction will cause it to fall, possibly injuring the animal or spooking him and others into running off. The high-line area should be a level spot and have some shelter for the animals. Shade during the daytime will usually offer some shelter in a rainstorm.

7-70. In desert environments where no trees are available, the same high line can also be used at ground level. The high line can be anchored with wooden stakes or a screw-in-type ground anchor at each end and one in the center. The animals will need to be checked on more frequently if this system is used to ensure the animals do not get entangled in their lead line.

Figure 7-15. High Line

7-71. The handler puts up the high line by starting with the lash cinch and going around an average-sized tree approximately 8 feet off the ground. He hooks the lash cinch into the rope and, if possible, runs the line so another tree is approximately in the center of the line. He ties a high-line knot approximately every 8 to 10 feet until he comes to the first tree. The handler goes around this tree and ties high-line knots until he gets approximately 2 feet from the tree that he intends to tie off. He uses the last knot to set up a dutchman, then goes around the tree and pulls the whole line as tight as possible. The high line can be reached by throwing the lead line over the top of the high line and pulling the high line down to secure the animal's lead line to the high line. Other factors to consider are as follows:

- When tying to a high line, stagger the animals tied to it—one on one side and then one on the other side.
- The high line can also make a good line to dry saddle pads when they are not being used for the animals.

PICKETING

7-72. Picketing is tying an animal up by a line attached to either a stake or a heavy log. It allows the animal a certain freedom of movement to walk around, get to grass, and perhaps drink water but does not give him the chance to wander off.

7-73. The handler can picket an animal in several ways. He normally uses either a halter and picket line or a hobble and picket line off one front leg. Depending on the animal's experience with the picket line, picketing by a leg is the safest. An animal picketed with a halter will sometimes hook a shoe in the halter while scratching his ear with a hind leg. If the picketed animal hooks the shoe of his hind leg in the picket halter, he usually falls down. If not found soon after falling, he may just die in this position.

7-74. The anchor end of the picket line can be either a good dry stake driven into solid ground or a drag log. Picket stakes should be stout and not brittle. When placing the stakes, the handler makes sure there is plenty of room between the stakes so two picket lines cannot cross. The drag log, when used, should be heavy enough so the animal picketed to it cannot move it, yet light enough so, if need be, it can be pulled to another picket area using a saddle animal and a line off the saddle horn.

7-75. A small bowline knot should be used in tying off a picket line to the stake. It should be down at ground level on the stake. This level will let the line turn around the stake and still not turn the stake and loosen it in the ground. An animal that pulls a stake and runs loose with the line is a danger to himself and the mission. Should the animal get into the timber and hang up the stake and line, it could cause a day's delay. Should the animal not be found, it most likely will die a slow death of thirst and starvation.

7-76. The handler should consider several things when picketing by any method. He should determine how much feed is in the picket area and if there is any obstacle in the ring that the picket line can foul on, thus shortening the ring area. The best area is usually on a good, grassy creek bank where the animal can get both feed and water off the same picket line.

7-77. When selecting which animals will go on the picket line, the handler should pick out animals new to the outfit and those he knows to be leaders. He should also consider any renegades or loners who hole up by themselves when running loose. If there is a mare with a weaner colt at home, she needs to go on a picket or on the high line. Otherwise, after she has had time to fill up on grass, she just might decide to go home.

THE NIGHT HORSE

7-78. If some of the stock is running loose during the night, a night horse is a must. The night horse is one that is kept picketed near where its handler sleeps. The handler uses him for emergencies during the night and for running down loose stock in the morning.

7-79. The night horse does not come in any particular color, size, or shape. It could easily be the most inefficient looking animal in the outfit. What counts is what is between its ears. Many an animal used as a pack animal when on the trail, though not considered a riding horse, might make a good night horse. The only way to come up with a top night horse is to try them all at different times. A good night horse should have several good points. He should—

- Be trusted to graze and fill up during the night and not just run in circles and whinny at the other loose animals all night.
- Handle well while running loose animals. One of the best reining animals in the whole outfit may develop a lot of bad habits, such as trying to buck its rider off when with loose animals. Worse yet, after

the loose animals are running, it may just stampede out of control with the rider, which is a very dangerous situation.

- Be easy to approach after dark and able to be ridden bareback. Quite often, should the loose animals start to leave during the night, it is possible to turn them very easily if the night horse is right there at the time. If the animal cannot be ridden bareback, the time it takes to saddle up might make it too late to easily turn the loose animals.

7-80. Many night horses develop the habit of watching where the loose animals go at night. Their hearing is far superior to humans. Often, while wrangling with a good night horse, it will want to head in a direction that the rider believes is wrong. The rider should trust the horse to help lead him to the loose animals. Quite often a night horse will want rein so he can smell the ground. Here again, its sense of smell is far better than the rider's and is being used to help locate the other animals. However, the horse is not totally in control; the rider most certainly will want to be reading tracks along with using the night horse's senses. Between the horse and the rider, loose animals are usually located quickly.

7-81. The night horse is often required to do quite a bit of running in bringing in the loose stock. This is bound to heat up the animal. Under such conditions, it must be cooled out. The most common mistake is for the rider to tie up the hot and lathered animal and head for the cook fire. In civilian pack outfits, many a supposedly experienced rider has been discharged for this one grave mistake. The best way to cool out a hot and lathered animal is to first unsaddle the animal. Using a currycomb and brush, the rider ruffles the hair over the back area. If he has some gunnysacks available, he should rub down the horse. Above all, the animal should **not** be fed grain or allowed any water until he is dry and cooled out. It could very possibly kill the animal or at the least cause a bellyache or colic. Walking the animal will also keep him from stiffening up. If the animal was good enough to keep on the night horse picket, he deserves the best of care when his job is done.

7-82. A night horse should be kept on the best grass and water and as close as possible to the sleeping area. A faint nicker from where the night horse is picketed may be the first sign that the other animals are moving out. The old mountain man may have slept with his rifle, but the handler should sleep with his bridle if he is the night horse rider.

TRANSPORTING SICK AND WOUNDED PERSONNEL

7-83. When confronted with an emergency, every situation varies according to what equipment is available. The unit leader should look over the horses to determine what can be used in a critical situation. There are many parts of saddles and gear strapped on them that can be pressed into service in ways other than those for which they were intended. This section is not meant to be an all-inclusive review of emergency procedures in the field. It is merely a brief look at things to consider and how they can be used expediently.

7-84. Using imagination and common sense and being observant of what is available to work with can help a person devise almost any type splint, bandage, sling, stretcher, or rig slings for transportation with equipment not necessarily intended for that purpose but which is at hand. However, using any part of the equipment may hinder the ability to ride out for help or transport the patient. Most of the items of equipment mentioned can be unbuckled or unsnapped rather than cut. If a strap needs to be cut, it should be cut close to a ring, buckle, or snap, so as to possibly save hours of needless repair later.

MOVEMENT

7-85. If the patient is subject to fainting or severe pain when moving, he should not be allowed to sit up in the saddle. If the person passes out and falls from the animal, the drop can be a meter or more. The fall will likely harm the patient and will upset the pack animal.

7-86. In most cases of back, neck, or spinal injury, it is best not to transfer the patient a great distance using horses. There is a lot to take into consideration, such as whether the patient is on an emergency backboard and how far it is to medical assistance. In most cases, where the patient will need to be transported only a short distance, a hand-held stretcher or travois pulled by a member of the team will be faster than rigging a horse for transportation. This way largely depends on how much help is available. There are several things that must **not** be done when transporting a victim with horses:

- Do not drape a wounded man head-down across a saddle.
- If the victim is inclined to pass out or is unsteady, do not let him ride by himself. Find out if the horse will ride double by sliding behind the saddle, without the victim. If the horse will ride double, ride behind the victim and hold him in the saddle.
- With a seriously sick or wounded patient, do not, in haste to get to help, trot or run the horse the victim is on. A slow, easy walk will get the patient to assistance in better condition.
- Do not at any time leave a horse unattended with a victim strapped or tied to the saddle. Maintain control of the horse's head at all times.

TRAVOIS OR PULKA LITTER

7-87. The difference between these two types of litters is in how the poles coming from the litter are attached to the pack animal. In the travois, the poles form an X over the back of the packsaddle. In the Pulka, the poles lie parallel along the sides of the animal's body. In most cases, the patient is below the level of the handler. This method makes most caregiving tasks easier since it is possible to reach over the patient without changing sides. The ends of the poles drag along the ground leaving ruts behind. This system does not work well if the ground is muddy or waterlogged. The patient will be soaked in only a few minutes. The handler must pick up the end of the litter and lift it over rocks, logs, or other obstructions. Crossing water obstacles requires the handler to hold up the litter and wade through the stream. Adrape stretching from the animal's rump to the poles of the litter will prevent gas or manure released from the animal to distress the patient (Figure 7-16, page 7-31).

Figure 7-16. Travois Litter

7-88. One of the simplest and fastest means of transporting a patient who cannot ride is a travois drawn behind one horse. This method requires getting a gentle horse used to pulling the travois before putting the patient on one for transportation. Personnel can construct a stretcher with two 15- to 20-foot strong poles that will support 200 pounds by running the poles through barrel hitches on the onside and offside of the animal. Again, it takes the horse a little while to get used to this rigging. If the emergency involves only two team members and the patient is conscious, the buddy should place the patient's head toward the horse and have him hold the reins. If this is not possible, he should tie the bridle reins to the poles, one on each side. This type of emergency stretcher is good for rough country or long-distance transportation. Most horses will tolerate the stretcher between them when they have a little time to become accustomed to the stretcher.

SUSPENDED LITTER

7-89. The suspended litter rides between two animals. It is long and requires a wide turning radius on any trail. The patient is about waist-high to the handler. It is easy to reach over and conduct any necessary task. The patient is kept high enough up to allow safe transport over rough ground, mud, low water, and other obstacles (Figure 7-17, page 7-32).

7-90. The primary problem with the suspended litter is that it requires two animals. If both are used to working together, only one handler is needed. If they march in stride, then the litter will rock from side to side. If they alternate legs, the litter may twitch. Both animals must keep the same speed and be very surefooted. The patient should be loaded with his head in the direction of travel. This way often helps prevent motion sickness.

FM 3-05.213

Figure 7-17. Suspended Litter

7-91. The handler can also stretch a large tarp or tent between two saddle horses that are side by side if he has a wide trail to travel. To rig the horses for this type of stretcher, it is wise to tie the horses' tails together to keep them from turning apart at the rear. A small pole tied between the bridle bits will keep them together at the head end. The tarp or tent should be pulled completely over the saddle and tied to the outside of the saddle. Two small poles, either run through or laced to the tarp, will make a stretcher in a very short time. In most cases, it is best to have the patient's head to the rear. Surprising as it may seem, this type of stretcher rides very comfortably.

STRETCHERS

7-92. The handler can use gunnysacks, canvas panniers, raincoats or saddle slickers, pannier tarps, saddle pads, mortuary adult shroud sheets (body bags), and blankets to make a stretcher. In most pack outfits, there are some gunnysacks used either to chink up a pannier or possibly to carry grain. By cutting two holes in the bottom corners of the sack and inserting two good poles, the handler has a first-class wilderness stretcher. Largely depending on what packhorses are present at the time, there is usually a set of deep canvas panniers. These canvas panniers have several emergency uses. The handler can make a stretcher with them, as with the gunnysacks. Another very important use is for lowering or raising an unconscious patient over such obstacles as a cliff or deep cut. The handler can cut two holes the size of the patient's legs in the bottom corners, slip his legs through these two holes, and pull the fabric up around his chest. The handler should test the knots used in tying a rope to the pannier with his own weight before trusting them on the patient. Raincoats, heavier coats, jackets, and saddle slickers also make a short distance stretcher by buttoning them up the front. Zippers are the best and most reliable. Two poles run through the inside of the garment and out the armholes make an emergency stretcher. Pannier tarps and large double saddle pads can make an emergency stretcher.

SPLINTS

7-93. The handler can use stirrup straps and fenders, pack pads, pannier tarps, pieces of wood, box panniers, and the bars out of packsaddles to make splints. There are many items of equipment, and often items loaded on a packhorse, that make an emergency splint. A backboard can be made by taking wood box panniers apart and either lashing or using the old nails in the box to nail or lash the flat boards to two small dry poles. The bars out of a packsaddle (using the flat surface) can also make a serviceable splint in a situation where other natural materials are not available. The stirrup straps and fenders out of a riding saddle are by far the best and simplest to obtain. The handler can unlace or unbuckle them and pull them out of the saddle without damaging the equipment. The stirrup straps and fenders from a riding saddle are already formed to the legs; therefore, using both straps and fenders can completely immobilize the legs of a victim. These work well also for the arms by using just one stirrup strap. Several small poles lashed flat and then padded with pack pads can make a backboard. An open rifle scabbard either down the seam or laid flat can make a splint.

STRAPS FOR LASHING TO BACKBOARDS

7-94. Latigos, lash cinches, saddle cinches, breast collars, brichens, and bridle reins can all be pressed into service. There are many large, heavy straps and cinches on a pack outfit or riding saddle. The handler can take off almost all of these very simply and make them serviceable for the bindings of splints or backboards. As mentioned before, he can also take off most all of these items without cutting them and thus damaging the equipment unnecessarily.

SMALL STRAPS TO BIND BANDAGES AND SPLINTS

7-95. Bridle reins, halter ropes, sling and lash ropes, saddle stirrups, and many of the straps on a packsaddle are useful as ties to hold things in place. There are many such straps on packsaddles and riding saddles. Bridle reins, for example, make a good long strap, and if needed for a long length of time, a person can still get by with one bridle rein. Almost all of the ropes, such as halter ropes and lash and sling ropes, can be unwound and one strand taken out while still leaving the rope serviceable. If small, tight bindings are required, the handler can use tail and mane hair from the horses. The handler should never use his or the patient's clothing to bind wounds and leave their bodies unprotected when there are so many other items available on the equipment. If the handler decides to leave the patient and ride for help, he can lay his saddle pads on the ground under the patient to help keep him warm and dry. Often there are two pads on a horse and, in an emergency, a rider could possibly get by with just one.

DIRECT PRESSURE IN SEVERE BLEEDING

7-96. Severe bleeding is always an emergency. If this situation arises and dressings are not readily available, the handler uses the sheepskin lining on the underside of the saddle to stop the blood flow. A small pad of this sheepskin with a clean dressing between it and the wound will in most instances pass for many layers of other dressings. A cool-back saddle pad also makes a good emergency compress for severe bleeding.

Chapter 8

Organization and Movement

Certain combinations of adverse weather, thick vegetation, and harsh terrain deny the use of wheeled or tracked vehicles in either a combat or logistics role. Mountainous terrain often restricts operations to those conducted by foot infantry. Heavily wooded areas, especially when associated with steep grades, have the same effect. Swamps, jungle-like vegetation, and certain types of cultivation may restrict the use of vehicles in lowlands. Weather, in combination with unfavorable terrain, may also deny or greatly restrict the use of aircraft in a combat or logistics role. In such situations, the commander who can move his troops, weapons, supplies, and equipment with the greatest speed and facility has a distinct advantage. Properly organizing, training, and equipping a combat pack animal detachment can give a commander this advantage.

ORGANIZATION

8-1. The pack animal detachment is usually the smallest fighting element. It can be a section, squad, or team of 10 to 20 individuals. The number of animals required to support these elements depends upon both the TOE and mission requirements.

8-2. The commander may task-organize the detachment squads or sections according to needs and requirements. For example, an 81-mm mortar section would not need as many individuals and animals to transport it as would a 107-mm mortar section. The mission will also have an impact on the size of the section; a raid with an 81-mm section will use fewer animals than sustained operations using the same weapons system. The squad or section requirements may vary from as few as five animals to as many as a dozen.

8-3. Other factors bearing on the organization of a pack detachment are load weight and the size of the items to be carried. The greater the total weight of the load, the more animals needed to carry it. If the items are large, even though lightweight, it will still take more animals.

DUTIES AND RESPONSIBILITIES

8-4. A pack animal detachment has unique duties and responsibilities. Most of these are leadership-related, and some are skills common to all.

8-5. The **train commander** is the commissioned officer or senior noncommissioned officer in charge of the pack train. He oversees the training, operation, and administration of the unit.

8-6. The **pack master** should be the platoon or team sergeant and the most knowledgeable about packing. He provides for the presence, care, and

maintenance of all pack equipment and the animals in the unit. He rides the entire column to check all loads and to observe the condition of the individuals and animals. The pack master also—
- Trains personnel in the proper methods of packing, to include saddling, adjusting equipment, balancing loads, and tying of standard hitches.
- Trains personnel in the proper care of animals and maintenance of pack equipment.
- Ensures maximum unit effectiveness through daily inspection of pack animals for injury.
- Supervises packing, conducts the march, maintains the animals and equipment, and disciplines the soldiers.
- Inspects loads, makes sure the animals on the march are not injured by shifting loads or saddles, and ensures prompt correction of deficiencies.
- Inspects and directs prompt repair of pack saddlery.

8-7. The **cargadores**, usually squad or section leaders, assist the pack master in all his duties and are qualified to perform the duties of the pack master in his absence. In addition, cargadores must be able to make all repairs normally made by the unit saddler. The cargadores also—
- Assign a load to each pack animal, ensuring the loads are balanced.
- Assign pack equipment, loads, and animals to the packers.
- Maintain order and discipline among the packers and ensure quiet and gentle treatment of the animals.
- Select areas for cargo piles and rig line in bivouac.
- Ensure proper care of pack equipment.
- Keep a memorandum of all cargo and equipment under their care, marking and tagging it if necessary.

8-8. The **packers** must train and care for both pack and riding animals. In the field, their duties include the maintenance, adjustment, and use of pack and riding saddlery and associated equipment. In addition, they prepare the cargo for packing and sling and lash loads using a variety of hitches. All detachment personnel should be packer-qualified.

MOVEMENT PROCEDURES

8-9. The pack detachment begins movement soon after the pack animals are loaded. Packers mount their riding animals. Two detailed riders move ahead of the train to contain any animals that are out of control. The detachment moves out in the march order directed by the commander in pack strings of no more than five animals. One rider leads each string and another follows it. The rider should always lead the strings from the onside of the pack animals (the string will be on the rider's offside). Normally, the detachment moves in file with the riders keeping the pack strings closed up to ensure communication between strings and to maintain column integrity. However, terrain and the probability of enemy contact from ground or air dictates the distance between strings or squads. When moving during hours of limited visibility, the rider should keep the column close to facilitate control. Troops riding on the flanks and rear of the strings make frequent counts of the animals to ensure against strays. Halts should be made as necessary to inspect and adjust loads and saddles. The commander, pack master, and cargadores make frequent inspections of the detachment en route, whether moving during daylight or at night.

8-10. Operating the pack animal detachment with individually led animals is considered uneconomical in terms of personnel. However, there are certain circumstances under which other factors are more important. Members of the weapons crews lead the animals that are assigned to combat units for the transport of heavy weapons and accompanying ammunition. Animals used in proximity to known hostile action areas are usually led individually to take advantage of all available cover and concealment. Evacuating casualties to aid stations by animal transport also requires this method of operation to ensure the easiest ride for the casualty, as well as to take maximum advantage of existing cover and concealment. Although training time for the operation may be reduced, the following tasks need to be followed:

- Train personnel to pack both lashed and hanger loads. Although the animal will normally proceed at a walk, with the rate of march seldom exceeding 3 1/2 miles per hour, make sure the loads are balanced and securely tied.
- Ensure the pack driver (person leading the animal) only exerts sufficient control over the lead animal to maintain its position on the trail or in the column. He should not interfere with the animal's freedom of movement or balance. A clumsy driver can cause a loaded animal to fall.
- Train the driver to lead the animals from the onside. He should guide the animal with his right hand grasping the lead rope, with the remainder of the lead rope in the left hand. He should give the animal enough slack so there is room enough for the animal to walk without stepping on him. If there is noise, danger, or confusion on the offside, the driver may reverse his position to offer some protection to the animal and quiet it.
- Train personnel to regulate the speed of the animal by the gentle, but effective, use of the lead rope. In column, always maintain the prescribed interval to prevent accordion action and undue fatigue in the rear elements.
- Teach the driver to counter the tendency of the animal to trot down slopes or jump over obstacles so he can maintain the normal rate of march and prevent displacement of the load.
- When leading an animal up steep slopes or over very rugged terrain, make sure the driver precedes the animal with about 3 feet of loose lead rope so that the animal may pick his footing. If the terrain is very rough or steep and he falls behind, it may be best to drop the lead rope and let the animal go. He can then catch the animal after the obstacle is passed.

FM 3-05.213

> **CAUTION**
> Under no circumstances will personnel hold the saddle breeching or the animal's tail to assist them in climbing.

STREAM CROSSING AT FORDS

8-11. Pack units will frequently come across streams or bodies of water where no bridges exist. These challenges, particularly in the spring or after a heavy rainfall, can be treacherous and demand special skills. Even small streams can cause problems when the water is rolling and muddy. All crossings should be checked carefully. If there is any doubt about a string's ability to cross easily, a single rider on a surefooted horse should be sent to check depth, bottom characteristics, and current strength. There is a huge difference in footing between a sandy or gravelly bottom and a bottom of large, mossy boulders. Similarly, fast water, deeper than a horse's belly, exerts its force against a much wider surface than water swirling around a horse's thin legs (Figure 8-1).

Figure 8-1. Stream Crossing

8-12. Although horses and mules can ford fairly deep water and are generally good swimmers, the crossing of even fordable water requires care and good judgment. Unit training must include accepted methods and techniques of stream crossing. Units should conduct ford reconnaissance before attempting to cross with a loaded pack unit. The load makes the animal somewhat top-heavy. The unit should try to cross streams with the animals moving against the current (upstream). The loads in combination with swift current, water deep enough to bear against the animal's body, and poor footing may cause the animal to lose its balance, fall, and drown. Under such circumstances, the unit should unload the animals and either lead or herd them across the body of water. The unit can also ferry loads across in the same manner as described in the following paragraphs. Table 8-1, pages 8-5 through 8-9, provides a water-crossing training schedule.

Table 8-1. Water-Crossing Training Schedule

	Time	Subject
Day 1	0730–0930	Instruction in swimming, general rules and adjustment of equipment. Risk assessment reviewed and read to all involved with training.
	0930–1130	Entering the water. Wading and practice in adjusting the equipment.
	1130–1200	Care of animals and equipment.
	1200–1300	Feeding animals and lunch.
	1300–1430	Turning animals out to graze. Waterproof packing of equipment.
	1430–1500	Coffee break.
	1500–1700	Repeat of wading practice.
	1700–1800	Care of animals and equipment.
	1800	Dinner.
	Time	**Subject**
Day 2	0730–0930	Wading and equipment adjustment.
	0930–1000	Care of animals and equipment.
	1000–1030	Coffee break.
	1030–1200	Construction of bridges or rafts. Animals grazing.
	1200–1300	Feeding animals and lunch.
	1300–1400	Construction of bridges and rafts. Animals grazing.
	1400–1500	Towing animals across water.
	1500–1600	Care of animals and equipment
	1600–1630	Coffee break.
	1630–1730	Water obstacle reconnaissance and analysis.
	1730	Dinner.

Table 8-1. Water-Crossing Training Schedule (Continued)

Day	Time	Subject
Day 3	0730–0930	Towing animals across water.
	0930–1030	Care of animals and equipment.
	1030–1100	Coffee break.
	1100–1200	Construction of bridges and rafts. Animals grazing.
	1200–1300	Feeding animals and lunch.
	1300–1400	Construction of bridges and rafts. Animals grazing.
	1400–1530	Swimming bareback.
	1530–1630	Care of animals and equipment.
	1630–1700	Coffee break.
	1700–1800	Water obstacle analysis.
	1800	Dinner.
Day 4	0730–0930	Swimming bareback.
	0930–1000	Care of animals and equipment.
	1000–1030	Coffee break.
	1030–1130	Care of animals in the field.
	1130–1230	Feeding animals and lunch.
	1230–1400	Construction of bridges and rafts. Animals grazing.
	1400–1600	Swimming bareback.
	1600–1630	Care of animals and equipment.
	1630–1700	Coffee break.
	1700–1800	Correction of deficiencies.
	1800	Dinner.
Day 5	0730–0930	Instruction in swimming with saddles.
	0930–1000	Care of animals and equipment
	1000–1030	Coffee break.
	1030–1200	Construction of bridges and rafts. Animals grazing.
	1200–1300	Feeding animals and lunch.
	1300–1430	Medical treatment of swimming injuries.

Table 8-1. Water-Crossing Training Schedule (Continued)

Day	Time	Subject
Day 5 (Continued)	1430–1600	Swimming with saddles.
	1600–1630	Care of animals and equipment.
	1630–1700	Coffee break.
	1700–1800	Medical treatment of injuries.
	1800	Dinner.
Day 6	0730–0830	Instruction in swimming with full pack.
	0830–1030	Swimming with full packs.
	1030–1100	Care of animals and equipment.
	1100–1130	Coffee break.
	1130–1230	Medical treatment of injuries.
	1230–1330	Feeding animals and lunch.
	1330–1430	Terrain analysis and exercise.
	1430–1600	Swimming with full packs.
	1600–1630	Care of animals and equipment.
	1630–1800	Correction of deficiencies.
	1800	Dinner.
Day 7	0730–0930	Swimming with full packs.
	0930–1030	Care of animals and equipment.
	1030–1100	Coffee break.
	1100–1130	Care of animals and equipment.
	1130–1230	Feeding animals and lunch.
	1230–1400	Rescue procedures.
	1400–1600	Swimming with full packs.
	1600–1630	Care of animals and equipment.
	1630–1700	Coffee break.
	1700–1800	Rescue procedures.
	1800	Dinner.

Table 8-1. Water-Crossing Training Schedule (Continued)

	Time	Subject
Day 8	0730–0930	Instruction in herd swimming.
	0930–1000	Care of animals and equipment.
	1000–1030	Coffee break.
	1030–1130	Terrain problems and terrain ride.
	1130–1230	Feeding animals and lunch.
	1230–1400	Team training.
	1400–1600	Herd swimming.
	1600–1630	Care of animals and equipment.
	1630–1700	Coffee break.
	1700–1800	Terrain problems and terrain ride.
	1800	Dinner.

	Time	Subject
Day 9	0730–0930	Herd swimming.
	0930–1000	Correction of deficiencies.
	1000–1030	Coffee break.
	1030–1100	Care of animals and equipment.
	1100–1200	Feeding animals and lunch.
	1200–1400	Instruction in special water-crossing problems, such as night and cold water.
	1400–1600	Swimming with full packs.
	1600–1630	Coffee break.
	1630–1730	Correction of deficiencies.
	1730–1800	Care of animals and equipment.
	1800	Dinner.

	Time	Subject
Day 10	0730–1000	Water-crossing problem.
	1000–1030	Coffee break.
	1030–1130	Correction of deficiencies.
	1130–1200	Care of animals and equipment.
	1200–1300	Feeding animals and lunch.
	1300–1400	Special problems in water crossing.

Table 8-1. Water-Crossing Training Schedule (Continued)

	Time	Subject
Day 10 (Continued)	1400–1600	Water-crossing exercise.
	1600–1630	Care of animals and equipment.
	1630–1700	Coffee break.
	1700–1800	Correction of deficiencies.
	1800	Dinner.

	Time	Subject
Day 11	0800–1600	Reserved for correction of deficiencies.

CROSSING UNFORDABLE WATER

8-13. Though an animal might be physically capable of swimming under a load, it upsets the animal's natural balance. When the pack unit must cross unfordable water, selected personnel swimming their riding animals should cross first to secure the farside, select the landing site on the far bank, and secure the hauling system for the pack animals. The nearside personnel should have animals ready to enter the body of water, ensure the hauling system is secure, and the pack animals' lead line is securely attached to the hauling system. The farside personnel should pull the pack animals and equipment to the far shore, disconnect lead lines from the hauling system, and secure the pack animals on the far bank. If the crossing is too wide for a hauling system, the unit can build poncho rafts to ferry the equipment and swim the animals freely across the body of water (Figure 8-2, page 8-10).

NOTE: Horses, mules, and dogs may cease to be effective when full-body immersion occurs (Figure 8-3, page 8-10) at water temperatures of 40 degrees F or below. Hypothermia will probably be noticeable within a half hour and life expectancy will be in jeopardy in an hour and a half. It will reach 50 percent loss in two hours and all but an overly fat animal will be dead or otherwise useless in three hours. At 32 degrees F, hypothermia will be obvious in 15 minutes, with 50 percent loss in one hour and total loss in an hour and a half. All of this is somewhat subjective and depends on body fat, physical condition, amount of food in the last 24 hours, general overall condition, and how well the animals are treated. All of this information is based on full-body immersion.

WATER TRAINING

8-14. Training animals to cross bodies of water is one of the most important activities if water obstacles are expected during an operation. The following procedures are recommended. All species can profit by this training, but it was originally developed for equines and may have to be modified for other animals.

FM 3-05.213

Figure 8-2. Crossing Unfordable Water

Figure 8-3. Full-Body Immersion

8-15. For safety, everyone involved should be questioned to see if they can swim and how well. The people who claim to be strong swimmers should be tested, and the best of them used as lifeguards. Good swimmers can also be used for mentoring beginners. Everyone must master survival floating.

8-16. The site of the training should be a quiet stretch of river or a small lake. Personnel should always check for quicksand or bad footing. The banks should not be steep or slippery. Many types of clay deposits are dangerous when wetted by splashing or tracked in water. Most slopes become soft after heavy traffic. The shore should be wide with a gradual shelving into the water. The bottom should have firm footing.

8-17. Each stage of training should be built on what precedes it. It is better to take longer than the times indicated here than to rush through training. Haste will only confuse the people and frighten the animals. Training should occur in groups of compatible people and animals. Peer pressure among the people and the herd instinct in the animals will help with instruction. Water training can be followed by a play period to further encourage both people and animals to look upon the water as a pleasant experience.

8-18. The unit should always use lifeguards and rescue equipment to prevent accidents. Unnecessary injuries will discourage students. Dangers should never be emphasized, but personnel should always be aware of the safety situation. Both animals and people lose body heat quickly even in warm water on a hot day. Personnel should watch for signs of chilling, such as shivering, and treat immediately.

TRAINING PROCEDURES

8-19. A 7/16-cm-thick rope should be stretched across the swimming area on the downstream side. Brightly colored floats should be attached at regular intervals. If possible, two motorboats or rowboats should be stationed at each side of the swimming area.

8-20. Poor or weak swimmers should wear life jackets during training. Voice amplification equipment should be available but used only in an emergency. Yelling only confuses the people and animals in the water.

8-21. A tent with cots, blankets, and a medic should be set up to care for anyone needing medical attention. Radio equipment should be operating at all times if an emergency occurs. Lifeguards should be posted at the entry and exit points. All lifeguards must be qualified in cardiopulmonary resuscitation (CPR). Two people should be assigned to catch any animals that break away from their riders.

ENTERING THE WATER

8-22. This phase is the most important part of training. Bad experiences at this stage can traumatize both humans and animals, resulting in problems for years to come. Patience on the part of the instructor can repay itself by reliable performance on the trail. The unit should use the madrina and mentors as examples.

8-23. Both humans and animals should be praised as each team moves deeper into the water. The animals can be rewarded with treats, which will produce a positive association with swimming and make training easier.

8-24. Wading should progress into deeper and deeper water until the animal must swim several strokes. At the same time, the animal must be required to turn in both directions while swimming.

8-25. Swimming animals move their legs in the same manner as walking but with exaggerated action. Most propulsion is from the forelegs. The hind legs keep the hindquarters afloat but they also reach further up than normal, which can cause them to become snagged in any straps or ropes trailing along. They can also kick a long-legged rider.

8-26. As the animal enters the water, it continues to keep contact with the bottom with all four legs as long as possible. An animal will remain calm even if the forehand begins to float as long as the hind legs are still solid.

8-27. The position of the rider significantly affects the balance of the animal. If the rider is too far forward, the forehand will sink and the head can go completely under. If the weight is too far back, the haunches go down and the forehand comes up. This impedes forward progress.

8-28. To get the animal to begin to swim, he must be ridden boldly into the water. Hesitation can cause balking. The rider sits well back to allow the forehand to float as soon as possible. As soon as the animal begins to struggle to find footing, the rider should quickly pull well up and force the hindquarters up. At this point, the animal floats and must begin to swim. The rider should keep the animal on a straight course, or as close as possible to straight, by aiming for the landing point on the opposite bank.

8-29. At no time should the animal be allowed to turn back to shore on its own. This movement will allow the animal to develop the conviction that it can refuse to cross. The first display of this behavior must be corrected and the animal required to swim properly. After that the trainer should be alert for subsequent attempts and make sure they are dealt with properly.

TOWING

8-30. If the unit decides to tow, it will need three ropes long enough to cross the body of water. Using a hauling system will require snap links. The unit will need the pack animal with packsaddles and the load. Animals must be haltered with a lead line attached to the chin ring.

8-31. The unit constructs a high line across the river attaching one end of the rope to a nearside tree and the other end to a farside tree. The high line should be at a height that allows the animal's head to be in a natural position when the lead line is attached. The handler places a snap link over the high-line rope. He attaches the nearside and farside hauling ropes to the snap link. Both ends of the hauling system ropes should be held by the pullers on both banks. Nearside personnel will attach the animal's lead line to the snap link using the figure 4 knot and move the animal to the water's edge. When nearside personnel signal they are ready, the farside hauler will begin to pull the animal into the water. The farside hauler must keep up with the animal's swimming speed to keep the rope from getting entangled around the animal's legs and cause drowning. When the animal reaches the farside bank, the hauler will disconnect him from the snap link, and secure him on the farside bank. The nearside hauler will pull the hauling system back across the body of water and attach the next animal for crossing (Figure 8-4).

Figure 8-4. Hauling System

8-32. If more than one animal is to be hauled across the body of water, the hauler attaches only the lead animal's lead line to the snap link. All other animals will be tied into the lead animal using the pigtails on the packsaddles. To avoid having animals lingering in the water at the far bank, the hauler moves the lead animal far enough up the bank so the other animals can follow.

NOTE: The unit should haul only as many animals as it can handle on the farside bank.

8-33. The rider or pack animals should be allowed to swim at the animal's own pace and not be pulled across. Personnel on the opposite bank should take up the slack as the team approaches. A person should never put a foot in loose coils of rope since any sudden movement outward can drag him into and under the water.

SWIMMING BAREBACK

8-34. When swimming bareback, equines should have bridles with snaffle bits and no other gear. Neck ropes and breast straps should be avoided since animals will fight them in the water. A halter should be used for other species.

8-35. Each animal should be ridden straight into the water bareback and required to keep a straight course. The rider slips off on the upstream side and swims alongside the animal. If the person is a weak swimmer, he can hold onto the pommel or crosspiece of the packsaddle. He should never pull so hard that the saddle is displaced.

8-36. The rider should not hang onto the reins since they may be needed to steer the animal. He must never hang onto the tail since it can be dislocated, broken, or even torn at the roots. The mane can be used in an emergency. The rider should also be careful not to get stepped on or kicked when the animal enters shallow water on the opposite bank. The rider should stay at least two meters apart from the nearest animal or a kick can result.

8-37. The rider should never attempt to change an animal's direction abruptly. He should never swim an animal in a circle if it gets ahead of its group. At the first sign of trouble or the animal's head begins to go under water, the rider must let go and get clear. The animal may thrash or roll over and a serious kick can result.

NOTE: For **minor** adjustments, the following procedures can be used:
- If the animal attempts to turn downstream or away from the rider, the rider should *gently* pull the upstream rein and force the animal to turn its head toward him.
- If the animal turns toward the rider and splashes water in his face, the rider should *gently* pull the downstream rein and turn the animal's head away. A sudden pull back on the reins can cause the animal to tumble over backward.

SWIMMING A HERD

8-38. The unit can swim animals either individually or in groups. Herd crossing is especially useful with strong currents. A mixed horses and mules group will cross more readily than when separated. Strong swimmers should take the natural leaders and start off immediately. If a madrina is used, she should cross first with the bell on. When she reaches the bank, the bell should be rung to let the others know where she is. It has also been noted that the animals can be called from the opposite bank by people imitating the appropriate call.

8-39. The unit should cut a ramp into the bank that will allow no more than two animals abreast to pass. Animals are held in a two-abreast line with handlers on the left side, and everyone facing the water. When the crossing is about to begin, personnel hold bamboo or thin limbs to form a rail corral with an opening at the chute.

8-40. Once the leaders begin to cross, the rest will follow them into the water. The rails can be closed in behind to encourage them to move but they must not be crowded. Fighting and kicking of other animals can result. Handlers catch the animals as they exit on the opposite shore and bring them back to a picket line.

SWIMMING ANIMALS WITH STRIPPED PACK OR RIDING SADDLES

8-41. Personnel should use girth straps that are one or more holes looser than normal. A person should be able to get two hands, laid flat (palm to palm), between the girth and the belly. The blanket will swell as it soaks. Equines should wear a snaffle bit with the reins unfastened. Other species should wear a halter.

8-42. The animals should be ridden into the water on a short rein. Pack animals wear a set of blankets tied with a surcingle. The riders slip off on the upstream side and cross as previously trained.

SWIMMING SADDLE AND PACK ANIMALS TOGETHER

8-43. Pack animals should have increased loads placed on the saddle as training progresses. If problems develop, excessive loads should be reduced during the training period. Complete saddles, bridles, and halters should be worn, but neck, breast, and chest straps should be removed. These straps seem to panic many species and cause them to drown.

8-44. Riders lead the assigned pack animals into the water. At first, pack loads should consist of soft, lightweight materials that do not soak up water too quickly. As time goes on, the loads should reflect the cargo that will be carried.

BIVOUAC

8-45. The pack animal unit should select its camps based on the results of both map and ground reconnaissance, if feasible. The selection depends on the requirements for the safety, health, and comfort of individuals and animals and the operational plans of the unit. The night camp should be reached just before twilight. Guards around the camp and grazing stock aid in preventing surprise attacks. If the camp is found, it can be vacated before a major assault can be launched. The bivouac should have little or no tentage unless conditions (freezing rain, below freezing temperatures) require it. As little area disturbance as possible should be made. Trees should not be cut down or holes dug except for pit latrines and garbage disposal. In hostile environments, concealment from air or ground observation and as much cover as possible are essential. The first consideration is always security, but unless the unit is carrying all the feed the pack and saddle animals will need, the next concern should be grass and water for the stock. It is a grave mistake to sacrifice grass, water, wood, and shelter for anything but security considerations. In selecting the grass, it should allow the stock to feed all night. Just because a meadow looks green does not mean it has plenty of grass in it. Often in high, alpine meadows, most of the vegetation consists of flowers and weeds that a horse will not eat. If the unit is unfortunate enough to camp in such a meadow, the stock will be weakened from lack of grass the next day on the trail. It is very irritating to ride or be leading an animal that insists on trying to graze as it is going down the trail. After security, the next concern should always be the care of the animals. Reconnaissance personnel should next look for the best footing for animals available in the area. Picket lines, high lines or, in more secure and extended situations, temporary corrals should offer level standing and good drainage with little possibility of flooding in sudden rains. The selected area should be large enough to provide adequate dispersion and be free from briers, debris, and poisonous plants. An area convenient to the route of march offers additional operational advantages.

8-46. The detachment or train commander, accompanied by selected personnel, precedes the train into camp and selects areas for the rigging line, high line, and the cargo so that the animals and equipment will be arranged systematically and readily available night or day. Upon arrival in camp, the pack train conducts the following procedures:

- Select personnel designate the cooking and sleeping areas, keeping in mind that they will always be located upstream from the animals.
- Upon arrival of the train, all personnel (except for the wranglers) dismount, secure their riding animals, unload the cargo, and slack off the cinches. Personnel unpack all animals before the first is unsaddled, and coil the lash and sling ropes. The wranglers remain mounted and ready to stop any animals that try to bolt from the camp.
- Personnel secure the animals as they are unloaded.
- When all loads are removed, the train commander gives the command to unsaddle.
- If conditions are favorable, handlers turn the animals out to graze or picket so they may roll and relax. If not, the animals stay on the high line while personnel clean the equipment and improve the site.
- Handlers feed the animals 45 minutes after arrival and water them 1 hour after arrival. Personnel then prepare their equipment for the next day's movement.

Chapter 9

Tactical Considerations

A pack animal detachment during movement, regardless of its combat mission once it reaches its destination, is a logistic transportation element. This fact alone severely limits the detachment's tactical capabilities. Even though the detachment may be considered a highly mobile unit, the presence of pack animals precludes the capability of maneuver.

The mission of the detachment while moving is the safeguarding and delivery of the cargo to its destination, not to stand and fight. This point is not to say that all is lost tactically while moving; it just means that compensation has to be made for the lack of maneuver and concentrated fires.

SECURITY

9-1. The pack detachment configured for movement forms into pack strings and usually in column formation. This alignment presents a long, linear target for the enemy. The troops are dispersed, making it difficult to bring concentrated, effective fire to bear upon enemy contact. The diligent use of scouts and outriders for flank security, along with extreme caution, are needed to a greater extent than for dismounted troops to make up for maneuver limitations.

9-2. Two factors affect not using overwatch formations. The first is terrain. The primary reason for using pack animals is difficult terrain. This fact necessitates using column formations. Second, the act of bounding requires one element to remain static. Animals that are loaded, strung together, and not moving are an accident waiting to happen. The following paragraphs explain how the detachment should cross specific types of areas, and Appendix C illustrates the various formations that may be used.

9-3. When the scouts reach a linear danger area (roads, trails, streams), they conduct a thorough reconnaissance for the best possible crossing point and attempt to reach the farside. Once they establish farside security, the scout leader returns to the column to lead it to the crossing point. Once the column arrives at the crossing point, lead personnel take up farside security and free the scouts to continue.

9-4. Commanders should consider crossing open or large areas as dangerous. Their use must be negotiated accordingly. The scouts should provide as much information as possible about the area to the commander. With this information, the commander decides how best to negotiate the obstacle. Bypassing the area is by far the best method but is not always practical. However, in arriving at a decision about movement, the leader must also consider METT-TC.

9-5. Whether during long or short halts, lead personnel establish and maintain security using the same procedures as a dismounted combat unit. The difference is the terrain to be defended and whether the animals will be in or out of the perimeter. The following principles apply:

- Animals kept on a high line within the perimeter may offer a reduction to security personnel, but unique problems can arise if the position is attacked. Casualties may arise from animals becoming frightened and breaking loose or stampeding.
- Animals kept on picket lines outside of the perimeter require listening posts or observation posts placed at a greater distance. During an attack on the detachment, these animals are susceptible to both enemy and friendly fire. The chances are good that a large percentage of the animals will either flee or be wounded.

COVER AND CONCEALMENT

9-6. The pack detachment must continuously consider concealment options during movement. Admittedly, it is next to impossible to conceal the evidence (animal waste, disturbed vegetation, and tracks) that a pack unit has moved through an area. However, personnel can take the following precautions to avoid being observed while on the move:

- Avoid skylining.
- Stay well within the tree line (if any).
- Contour the terrain.
- Camouflage the loads.
- Avoid open areas, if possible, or cross them quickly.

9-7. During extended halts (when loads are unsaddled), personnel should use the proper camouflage. Cover for the animals, while desirable, may not be possible or practical. Personnel should be alert to animal noise, both vocal and from movement, and address the odors associated with animals. There are no clear-cut solutions to the problems of concealment. Personnel should use a common-sense approach and practice the following tips:

- Use as much natural cover and foliage as possible.
- Use camouflage nets.
- Apply proper field sanitation techniques, to include keeping animal waste policed.
- Control animal noise as much as possible. If the animals' vocal cords have not been cut, keep the animals quiet by maintaining a relaxed atmosphere.

ACTIONS ON CONTACT

9-8. As previously mentioned, a pack detachment organized for movement is not a maneuver unit. Therefore, it must act accordingly when contact is made.

9-9. When ambushed, those elements caught in the kill zone should quickly escape in any feasible direction. Forward elements, not in the kill zone, move in the direction of march as fast as possible. Those in the rear should move in the direction away from the kill zone, quickly but cautiously. All elements should make every effort to link up at a preplanned rally point. Elements not under direct fire should attempt to control and safeguard the animals.

9-10. Actions under indirect fire are handled in much the same way as an ambush. It may not be necessary to move to a rally point. The unit may stay out of harm's way until the barrage has been lifted and then continue in the direction of march. The commander states the actions for these events and incorporates them in the unit's SOP before movement.

9-11. The key to surviving an air attack is dispersion and continuous movement. The pack detachment always accounts for personnel, weapons, and equipment after it moves to the designated rally point, establishes security, and reestablishes the chain of command.

NOTE: Common sense, preparation, and good planning are the keys to surviving as a pack detachment in a hostile environment.

URBAN ENVIRONMENTS

9-12. Use of any kind of animals in an urban environment requires detailed planning and preparation. Even a clean, secure urban setting wears down an animal and its shoes rapidly. A damaged city with extensive rubble provides more danger in the form of sharp rock, broken glass, and other debris that can permanently damage animals. The best choice when planning to use pack animals to support an urban operation is to halt them at a secure location outside of any built-up areas and move the supplies in by foot. However, if the detachment must move the animals through a built-up area, it should try to avoid rubble, broken glass, nails, and other debris that will injure the animal.

9-13. A well-equipped farrier has special shoes for pack animals if the detachment is expected to stay in an urban environment for any period of time. Thick rubber horseshoes work well on pavement and concrete, but do not provide any traction on grass or any kind of dirt. Police departments sometimes use a complex shoe with a plastic plate to protect the inner hoof, a steel shoe over that, and carbide bits welded to the shoe. The carbide bits provide traction on asphalt and concrete, the shoe itself provides traction off-road, and the plastic protects the horse's inner hoof from nails, glass, and other debris.

9-14. Normally, the pack animal detachment will not have access to any special equipment to protect the animals. As mentioned earlier in this manual, shoes wear out much quicker on rocky or rough terrain. This is even more apparent when operating in any kind of urban environment. The detachment must check the animals' feet for any damage constantly throughout the day and must pay close attention to the animals during movement to watch for any injuries as soon as they occur. Routes must be selected and checked ahead of time to reduce the danger to the animals as they move. Special care must be taken to avoid any rubble, concrete, or asphalt along the route.

Chapter 10

Llamas and Other Animals

Although this manual focuses on horses and mules, this chapter provides basic information about other types of pack animals that a team may encounter. With the exception of llamas, these animals will normally come with a native handler that should be considered the primary source of information for the use of that animal.

LLAMAS

10-1. Llamas (Figure 10-1) are members of the camel (camelid) family. In addition to the well-known, one-humped dromedary camel of the Middle East and the two-humped Bactrian camel of Asia, there are four native members of the camel family in the Americas today—the llama, a domesticated beast of burden regarded throughout the world as the premier symbol of South American animals; the alpaca; the free-ranging guanaco; and the wild vicuna.

Figure 10-1. Llama

CHARACTERISTICS

10-2. While viewed in the pasture or close contact, all llamas have a striking beauty owing to their elegant wool and graceful posture. Llama wool ranges from white to black, with shades of gray, brown, red, and roan in between. Markings can be in a variety of patterns from solid to spotted. Little variation is found in guanacos or vicunas, which are light brown with white undersides.

10-3. Mature llamas weigh an average of 280 to 350 pounds, but range from 250 to 500 pounds. Full body size is reached by the fourth year and, while there are no obvious differences between the sexes, males tend to be slightly larger. They are long-lived, with a normal life span of 15 to 20 years.

10-4. Like cattle and sheep, llamas are multistomached and chew the cud. They have a hard upper gum (no upper teeth in front), grinding upper and lower molars in back, and a cleverly designed upper lip for grasping forage in unison with the lower incisors. Adult males develop large, sharp upper and lower canines (wolf teeth or fangs) for fighting. The unit's veterinarian may remove these teeth to prevent injury to males pastured together or to females being bred.

10-5. The llamas' unique, specially adapted foot makes them remarkably surefooted on a variety of terrain, including sandy soils and snow. It is two-toed with a broad, leathery pad on the bottom and curved nails in front. The small, oblong, bare patches on the side of each rear leg are not vestigial toes (chestnuts as found on horses), but metatarsal scent glands, suspected to be associated with the production of alarm pheromones. An additional scent gland is located between the toes.

HOUSING AND FENCING

10-6. Fencing can be woven wire, cattle wire panels, wooden rails or poles, chain link, or electric. Barbed wire is not recommended. The fences should be at least 4 feet high and dog-proof, if possible.

10-7. The llamas should have a three-sided shelter to provide shade and protection from extreme heat, cold, wind, and rain. If the area has severe chill factors in winter, a completely enclosed shed is necessary. Heat stress should be a concern if there are hot and especially humid summers. The unit may want to consider shearing or clipping the llamas for this type of weather. If the unit does not intend to shear the stock, then a sprinkler, wading pool, or small pond may be helpful in keeping the llamas cool for the summer.

10-8. If the animals are kept in a large pasture, a small 12- to 20-foot-square catch pen will make it easier to catch them. Feeding and watering troughs should be clean, high enough to be free of possible fecal contamination, and spacious enough to allow access by all animals. Fresh water should always be available.

TRANSPORTATION

10-9. Llamas are easy to transport and require no specialized equipment. A covered, windproof pickup, van, and horse or utility trailer with sufficient room for animals to stand comfortably works well. Good ventilation is important in both summer and winter. Straw makes excellent bedding in a windproof enclosure. Llamas should be provided hay for food and offered water free-choice at least every 6 hours depending on heat (it will spoil if left with the animals). Llamas normally lie down (kush) once the vehicle starts moving. When transporting babies and mothers on long hauls, the unit should stop periodically to allow for nursing.

CARE AND FEEDING

10-10. If familiar with the care of other domestic livestock, caring for llamas will be comparatively easy, with a minimum of veterinary assistance required. If uncertain of the health of a new animal, the handler should consider isolating it in sight of, but separate from, the others for the first 2 weeks to prevent accidental introduction of any illnesses and to allow for getting acquainted with the animal. The handler should make sure the llama is eating and ruminating, and eliminating pelleted feces.

10-11. If the unit has not already done so, the handler should locate a veterinarian in the operational area. If he is inexperienced with llamas, information and assistance is available through various animal clinics in Canada and the United States. The unit should have a veterinarian check the llama and take a fecal sample to determine if worming is necessary.

10-12. Llamas are amazingly hardy animals and have very few problems with disease. To ensure good health, the unit should establish a regular schedule for cleaning their dung piles and a preventative medicine program that may include protection from enterotoxemia, tetanus, leptospirosis, and internal and external parasites. Llamas should be dewormed at least every 6 months. The unit should also check with the veterinarian or agricultural extension agent to see if any vital trace elements or minerals are deficient or present in toxic amounts in the operational area.

10-13. Although llamas have long been arid land dwellers, they thrive in the wide array of temperate environments throughout Canada and the United States, including Alaska. They are highly adaptable feeders, being both grazers (grasses and forbs) and browsers (shrubs and trees).

10-14. Because of a relatively low protein requirement due to their efficient digestive systems, they can be kept on a variety of pastures or hay. They eat about 2 to 4 percent of their body weight in dry matter each day. Without pasture, a 100-pound bale of hay will last an adult llama around 10 days, much longer than the same amount would last an adult horse. If grazing the llamas, the handler should plan on about three to five animals per acre on a moderate-producing pasture.

10-15. When good hay is available, grain is recommended only for working pack animals and nursing females. Sheep mineral blocks and salt blocks (with selenium wherever necessary) should be available free-choice. Granulated minerals are somewhat more wasteful than mineral blocks, but are easier to eat. High-protein grain mixes prepared for other livestock should generally not be given to a healthy llama on a good diet, unless it is a female nursing or close to giving birth. Llamas are not prone to bloat, but have been known to do so if they get into a grain bin.

10-16. Llamas require less water than most domestic animals, but should have an unlimited, fresh, clean supply at all times. They tend to drink less in winter and when on lush, green pasture, and more when working or lactating, especially in summer.

10-17. Unless the llamas are pastured on hard or rocky ground, their toenails need trimming only once or twice a year. It is easy to do with horse hoof trimmers or sheep toenail nippers, but the handler should consult the veterinarian before his first attempt.

HABITS AND BEHAVIOR

10-18. Llamas have a dignified, aristocratic manner about them. Because of their curiosity, they have a delightful habit of coming close to sniff strangers. But despite your natural temptation to hug and cuddle them, they prefer not to be petted except on their necks and woolly backs. They are gentle, do not spook easily, and rarely bite or kick unless provoked. They are highly social animals and need the companionship of other llamas.

10-19. Llamas communicate their moods with vocalizations and a series of tail, body, and ear postures. Learning this llama language is one of the joys of ownership. Humming is a common manner of communication between llamas and indicates a variety of moods from contentedness to concern. Another interesting llama expression is the shrill, rhythmic alarm call emitted at the sight of a strange animal (especially coyotes and dogs) or a frightening situation.

10-20. Seldom directed at people, spitting is usually related to food disputes and to establish a pecking order between llamas. The exception to this is if a llama has been mishandled or becomes imprinted on people (through bottle-feeding as a baby). As with bottle-fed stallions, bulls, and rams, bottle-fed male llamas that have not been gelded at an early age can be potentially dangerous as adults (because they lack a normal fear of people and regard them as competitors). Therefore, males not intended for breeding and males that were bottle-fed must be gelded as early as possible to avoid undesirable behavior.

10-21. Llamas are remarkably clean and even large herds are quite odorless. Dung-piling behavior is an important means of spacial orientation and territorial marking for these historically open-habitat animals—a convenience when cleaning their pens. By taking advantage of this habit, the handler can encourage the animals to establish dung piles in a new pen by "probating" four to five sites per acre with a shovel full of llama dung. The llama may frequently roll in the dirt, taking a dust bath to help maintain a healthy, fluffy coat of wool.

USES AND TRAINING

10-22. "What are they used for?" is a question commonly asked of llama handlers. The domesticated llama in South America is used for low-fat meat, wool, hide, manure (for fertilizer and fuel), and as a beast of burden. This multipurpose animal lends itself well to the needs of the South American people. The llama in North America is used for investment, breeding, wool production, packing, tourism, sheep guards, shows, competitions, and as a source of diversification.

10-23. Llama wool production is a multimillion dollar industry in South America and makes llamas appealing to spinners and weavers there. Llamas have soft, fine wool for insulation against cold and heat. The llama's wool can also be harvested yearly by clipping or shearing and be made into beautiful garments and blankets. The yearly wool harvest produces 3 to 8 pounds of grease-free fleece with a fiber length of 4 to 7 inches. Year-round brushing yields about the same results and leaves the long, coarser guard hairs in place.

10-24. The llamas' ability as beasts of burden has been rediscovered by hikers, hunters, and forest work crews. Their hardiness, surefootedness, and common sense make them an excellent pack animal, trail companion, and even a great golf caddy. They are quiet, unobtrusive, and so easy to manage that children love to lead them. The llamas' agility allows them to negotiate terrain that would be difficult or impossible for traditional pack animals. Because of their padded feet and ability to browse, they have minimal impact on the backcountry. When confronted by other pack stock, unexpected situations, and sudden movement or noises, llamas remain calm and unruffled.

10-25. Males are most commonly used for packing and, depending on maturity, weight, and condition, will tote 50- to 120-pound packs 10 to 15 miles a day. A variety of packs and halters are available for llama use.

10-26. Because they are gentle and easy to train, llamas are popular attractions in parades, shows, fairs, and community events, and are fun to take to school, hospital, or nursing home visits.

10-27. Llamas are fast becoming the choice for protecting sheep flocks from predators such as stray dogs and coyotes. They are also good competitors. Conformation is judged at llama shows to determine the best of each halter class. The llama's training and disposition are judged in performance competitions such as obstacle and driving classes.

10-28. Surprisingly, there is a market for llama manure. It is high in nitrogen, basically odorless, weed-free, and easily collected. It is a valuable plant and garden fertilizer.

10-29. Their docile nature makes llamas extremely easy to train to accept a halter, lead, kush, carry a pack, load in and out of a vehicle, pull a cart, or carry a lightweight rider. With just a few repetitions, they will pick up and retain any of these skills. Llamas, with minimal training, are easy to handle when trimming nails, brushing or shearing, or when health problems necessitate touching them in sensitive places.

CAMELS

10-30. There are two basic types of camels (Figure 10-2, page 10-6). The dromedary (short fur, one hump) is found in Southwest Asia and weighs about 2,000 pounds. The Bactrian (long fur, two humps) is found in Central Asia and weighs about 3,000 pounds. A dromedary can carry packs weighing 600 to 700 pounds (35 percent of body weight) for 7 to 8 hours for a distance of 25 to 35 miles. Bactrians can carry a load of 900 to 1,000 pounds (35 percent of body weight) for 7 to 8 hours for a distance of 25 to 35 miles. Camels can also pull loads of 1,650 pounds or 3,300 pounds by two, per day at normal walking speed.

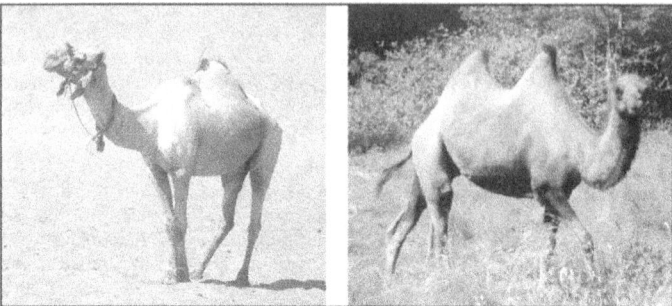

Figure 10-2. Two Types of Camels

10-31. Camels have broad, flat, leathery pads with two toes on each foot. When the camel places its foot on the ground, the pads spread, preventing the foot from sinking into the sand. When walking, the camel moves both feet on one side of its body, then both feet on the other. This gait suggests the rolling motion of a boat, explaining the camel's "ship of the desert" nickname. They need very little water and can travel for days without drinking at all. Camels are clumsy-looking, rather ugly animals, and have a lousy reputation because they are believed to spit and kick at people. This perception is not accurate because well-handled camels are safe to work with and be around.

10-32. As stated above, camels will come with native handlers to take care of them. The handler should be considered the detachment's first choice when planning operations using a camel.

DOGS

10-33. The use of dogs as an auxiliary in war is as old as war itself. Primitive man used dogs to guard his family, his belongings, and himself. He also took his dog into battle with him when rival tribes clashed. Throughout the history of warfare, dogs have gone into combat at the side of their masters or have been used in direct support of combat operations.

10-34. Dogs have long been recognized by the military as important for war and security purposes. The Army has used various dog breeds and has determined the German shepherd to be the most suitable for scouting because of its working ability, temperament, size, availability for procurement, and adaptability to all types of climate and terrain. These dogs are especially good at detecting ambushes. While "reading" the dog, the handler must prevent him from barking, growling, whining, or otherwise making a noise, which would be audible to a lurking enemy. Such a reaction on the dog's part might be fatal to both dog and handler in combat.

10-35. The dog is urged to move silently through built-up areas. After learning the basic rules, the dog is then trained to detect an enemy decoy planted farther and farther away. Lessons are repeated under varied conditions at different times of the day and night until the animal becomes expert enough to detect an enemy at distances up to 500 yards or more.

10-36. Dogs have many uses in combat and were valued especially by the infantryman. Taken with patrols into no-man's-land, their keen senses reduce the chance of fatal surprise or ambush by giving silent warning of a concealed enemy. During the hours of darkness or when visibility is poor, dogs can guard command or observation posts against enemy infiltration, a problem that was acute in Korea and Vietnam.

10-37. After the dog develops a strong fondness for two or more handlers, it is taught to run from one to the other. One handler releases the dog and commands the animal to "report." As soon as the dog reaches the second handler, it receives warm praise. Just before the dog runs, its "choke" chain is removed and a "messenger" collar is put on. Soon the dog learns to associate this collar with the job of running from one man to the other. As the lessons proceed, the distance between the men is increased beyond the range of the dog's vision. The dog now learns to trail its masters by scent. Frequent repetition and runs of varied distances over different kinds of terrain finally develop the animal's dependability as a messenger.

10-38. In training, the dog customarily carries a pack, which can be loaded with supplies or ammunition. The dog also learns to lay field telephone wire from a spool mounted on a specially constructed pulling harness. Before being considered fully trained, the dog must be able to follow a scent up to distances of 5 miles. It must be able to carry up to 30 pounds of ammunition and supplies over rough terrain. It must also demonstrate capability in carrying and stringing a 1-mile spool of telephone wire between two points. Using dogs as pack animals has advantages and disadvantages.

10-39. The advantages are that a dog—

- Has intelligence, making it a tool that can be used for tactical advantage.
- Has keen senses of sight, smell, and hearing.
- Can trail handlers by scent over great distances.
- Can sleep anywhere the handlers sleep.
- Can eat the same food as the handlers.
- Can carry packs up to 30 pounds.
- Can perform various duties; for example, string communications wire and carry messages.
- Shows early warning against possible enemy, mines, or booby traps.
- Can clear tunnels.
- Requires short periods of time to recover.
- Adapts well to changes in climate.
- Can pull sleds or small carts.
- Is a good swimmer.

10-40. The disadvantages are that a dog—
- Is a 24-hour-a-day responsibility of a Soldier.
- Adds additional stress to the handler by being an additional responsibility.
- Needs proper equipment to operate continuously in summer, winter, and in rugged terrain.
- Can have his performance adversely affected by extreme temperatures.
- Can be rendered completely ineffective by pad injuries.
- Needs additional food and water.
- Can be adversely affected by noxious odors.

ELEPHANTS

10-41. Elephants are considered an endangered species and as such should not be used by U.S. military personnel. There are about 600,000 African elephants and between 30,000 and 50,000 Asian elephants. Approximately 20 percent are in captivity, so it is difficult to estimate their numbers exactly. The Convention of International Trade in Endangered Species regards both species as threatened. Elephants are not the easygoing, kind, loving creatures that people believe them to be. They are, of course, not evil either. They simply follow their biological pattern, shaped by evolution. The secret of becoming a good trainer is to learn this pattern. The handler can then apply it to himself and the elephants under his control.

Appendix A

Weights, Measures, and Conversion Tables

Tables A-1 through A-5, pages A-1 and A-2, show metric units and their U.S. equivalents. Tables A-6 through A-15, pages A-2 through A-5, are conversion tables.

Table A-1. Linear Measure

Unit	Other Metric Equivalent	U.S. Equivalent
1 centimeter	10 millimeters	0.39 inch
1 decimeter	10 centimeters	3.94 inches
1 meter	10 decimeters	39.37 inches
1 decameter	10 meters	32.8 feet
1 hectometer	10 decameters	328.08 feet
1 kilometer	10 hectometers	3,280.8 feet

Table A-2. Liquid Measure

Unit	Other Metric Equivalent	U.S. Equivalent
1 centiliter	10 milliliters	0.34 fluid ounce
1 deciliter	10 centiliters	3.38 fluid ounces
1 liter	10 deciliters	33.81 fluid ounces
1 decaliter	10 liters	2.64 gallons
1 hectoliter	10 deciliters	26.42 gallons
1 kiloliter	10 hectoliters	264.18 gallons

Table A-3. Weight

Unit	Other Metric Equivalent	U.S. Equivalent
1 centigram	10 milligrams	0.15 grain
1 decigram	10 centigrams	1.54 grains
1 gram	10 decigrams	0.035 ounce
1 decagram	10 grams	0.35 ounce
1 hectogram	10 decigrams	3.52 ounces
1 kilogram	10 hectograms	2.2 pounds
1 quintal	100 kilograms	220.46 pounds
1 metric ton	10 quintals	1.1 short tons

Table A-4. Square Measure

Unit	Other Metric Equivalent	U.S. Equivalent
1 square centimeter	100 square millimeters	0.155 square inch
1 square decimeter	100 square centimeters	15.5 square inches
1 square meter (centaur)	100 square decimeters	10.76 square feet
1 square decameter (are)	100 square meters	1,076.4 square feet
1 square hectometer (hectare)	100 square decameters	2.47 acres
1 square kilometer	100 square hectometers	0.386 square mile

Table A-5. Cubic Measure

Unit	Other Metric Equivalent	U.S. Equivalent
1 cubic centimeter	1,000 cubic millimeters	0.06 cubic inch
1 cubic decimeter	1,000 cubic centimeters	61.02 cubic inches
1 cubic meter	1,000 cubic decimeters	35.31 cubic feet

Table A-6. Temperature

Convert From	Convert To
Fahrenheit	Celsius
	Subtract 32, multiply by 5, and divide by 9
Celsius	Fahrenheit
	Multiply by 9, divide by 5, and add 32

Table A-7. Approximate Conversion Factors

To Change	To	Multiply By	To Change	To	Multiply By
Inches	Centimeters	2.540	Ounce-inches	Newton-meters	0.007062
Feet	Meters	0.305	Centimeters	Inches	3.94
Yards	Meters	0.914	Meters	Feet	3.280
Miles	Kilometers	1.609	Meters	Yards	1.094
Square inches	Square centimeters	6.451	Kilometers	Miles	0.621
Square feet	Square meters	0.093	Square centimeters	Square inches	0.155
Square yards	Square meters	0.836	Square meters	Square feet	10.76
Square miles	Square kilometers	2.590	Square meters	Square yards	1.196
Acres	Square hectometers	0.405	Square kilometers	Square miles	0.386
Cubic feet	Cubic meters	0.028	Square hectometers	Acres	2.471

Table A-7. Approximate Conversion Factors (Continued)

To Change	To	Multiply By	To Change	To	Multiply By
Cubic yards	Cubic meters	0.765	Cubic meters	Cubic feet	35.315
Fluid ounces	Millimeters	29.573	Cubic meters	Cubic yards	1.308
Pints	Liters	0.473	Millimeters	Fluid ounces	0.034
Quarts	Liters	0.946	Liters	Pints	2.113
Gallons	Liters	3.785	Liters	Quarts	1.057
Ounces	Grams	28.349	Liters	Gallons	0.264
Pounds	Kilograms	0.454	Grams	Ounces	0.035
Short tons	Metric tons	0.907	Kilograms	Pounds	2.205
Pounds-feet	Newton-meters	1.356	Metric tons	Short tons	1.102
Pounds-inches	Newton-meters	0.11296	Nautical miles	Kilometers	1.852

Table A-8. Area

To Change	To	Multiply By	To Change	To	Multiply By
Square millimeters	Square inches	0.00155	Square inches	Square millimeters	645.16
Square centimeters	Square inches	9.155	Square inches	Square centimeters	6.452
Square meters	Square inches	1,550	Square inches	Square meters	0.00065
Square meters	Square feet	10.764	Square feet	Square meters	0.093
Square meters	Square yards	1.196	Square yards	Square meters	0.836
Square kilometers	Square miles	0.386	Square miles	Square kilometers	2.59

Table A-9. Volume

To Change	To	Multiply By	To Change	To	Multiply By
Cubic centimeters	Cubic inches	0.061	Cubic inches	Cubic centimeters	16.39
Cubic meters	Cubic feet	35.31	Cubic feet	Cubic meters	0.028
Cubic meters	Cubic yards	1.308	Cubic yards	Cubic meters	0.765
Liters	Cubic inches	61.02	Cubic inches	Liters	0.016
Liters	Cubic feet	0.035	Cubic feet	Liters	28.32

Table A-10. Capacity

To Change	To	Multiply By	To Change	To	Multiply By
Milliliters	Fluid drams	0.271	Fluid drams	Milliliters	3.697
Milliliters	Fluid ounces	0.034	Fluid ounces	Milliliters	29.57
Liters	Fluid ounces	33.81	Fluid ounces	Liters	0.030
Liters	Pints	2.113	Pints	Liters	0.473
Liters	Quarts	1.057	Quarts	Liters	0.946
Liters	Gallons	0.264	Liters	Gallons	3.785

Table A-11. Statute Miles to Kilometers and Nautical Miles

Statute Miles	Kilometers	Nautical Miles	Statute Miles	Kilometers	Nautical Miles
1	1.61	0.86	60	96.60	52.14
2	3.22	1.74	70	112.70	60.83
3	4.83	2.61	80	128.80	69.52
4	6.44	3.48	90	144.90	78.21
5	8.05	4.35	100	161.00	86.92
6	9.66	5.21	200	322.00	173.80
7	11.27	6.08	300	483.00	260.70
8	12.88	6.95	400	644.00	347.60
9	14.49	7.82	500	805.00	434.50
10	16.10	8.69	600	966.00	521.40
20	32.20	17.38	700	1127.00	608.30
30	48.30	26.07	800	1288.00	695.20
40	64.40	34.76	900	1449.00	782.10
50	80.50	43.45	1000	1610.00	869.00

Table A-12. Nautical Miles to Kilometers and Statute Miles

Nautical Miles	Kilometers	Statute Miles	Nautical Miles	Kilometers	Statute Miles
1	1.85	1.15	60	111.00	69.00
2	3.70	2.30	70	129.50	80.50
3	5.55	3.45	80	148.00	92.00
4	7.40	4.60	90	166.50	103.50
5	9.25	5.75	100	185.00	115.00
6	11.10	6.90	200	370.00	230.00
7	12.95	8.05	300	555.00	345.00
8	14.80	9.20	400	740.00	460.00
9	16.65	10.35	500	925.00	575.00
10	18.50	11.50	600	1110.00	690.00
20	37.00	23.00	700	1295.00	805.00
30	55.50	34.50	800	1480.00	920.00
40	74.00	46.00	900	1665.00	1033.00
50	92.50	57.50	1000	1850.00	1150.00

Table A-13. Kilometers to Statute and Nautical Miles

Kilometers	Statute Miles	Nautical Miles	Kilometers	Statute Miles	Nautical Miles
1	0.62	0.54	60	37.28	32.38
2	1.24	1.08	70	43.50	37.77
3	1.86	1.62	80	49.71	43.17
4	2.49	2.16	90	55.93	48.56
5	3.11	2.70	100	62.14	53.96

Table A-13. Kilometers to Statute and Nautical Miles (Continued)

Kilometers	Statute Miles	Nautical Miles	Kilometers	Statute Miles	Nautical Miles
6	3.73	3.24	200	124.28	107.92
7	4.35	3.78	300	186.42	161.88
8	4.97	4.32	400	248.56	215.84
9	5.59	4.86	500	310.70	269.80
10	6.21	5.40	600	372.84	323.76
20	12.43	10.79	700	434.98	377.72
30	18.64	16.19	800	497.12	431.68
40	24.86	21.58	900	559.26	485.64
50	31.07	26.98	1000	621.40	539.60

Table A-14. Yards to Meters

Yards	Meters	Yards	Meters	Yards	Meters
100	91	1000	914	1900	1737
200	183	1100	1006	2000	1828
300	274	1200	1097	3000	2742
400	366	1300	1189	4000	3656
500	457	1400	1280	5000	4570
600	549	1500	1372	6000	5484
700	640	1600	1463	7000	6398
800	732	1700	1554	8000	7212
900	823	1800	1646	9000	8226

Table A-15. Meters to Yards

Meters	Yards	Meters	Yards	Meters	Yards
100	109	1000	1094	1900	2078
200	219	1100	1203	2000	2188
300	328	1200	1312	3000	3282
400	437	1300	1422	4000	4376
500	547	1400	1531	5000	5470
600	656	1500	1640	6000	6564
700	766	1600	1750	7000	7658
800	875	1700	1860	8000	8752
900	984	1800	1969	9000	9846

Appendix B

21-Day Pack Animal Training Program

Hours	Day 1

1st	- Teach animals to be caught with ease. This practice can easily be accomplished by having the handler—
	  - Walk into the corral rattling a can with sweet grain.
	  - Call the name of the animal and offer feed.
	- Do this every day so that the animal associates the sound of its name with coming to the handler. (The animal also associates the rattle of the grain can with a reward.)
	- Check the Preston brands of all animals against the Animal Record Cards.
	- Record the name the handler has given the animal.
	- Check the equipment to be used with the remounts and place it in quarantine for the 21-day period.
	- Tie the animals to secure posts; do not use flimsy articles, such as the boards of a fence.

2d	- Select personnel to work with the remounts. - Assign one person to an animal. - Ensure that the best trainers work with the more nervous animals (those that spook easily).
3d	- Inspect all the animals for signs of shipping diseases. - Report all cases of nasal or watery eye discharge, ringworm, or any other signs of diseases. - Place any suspected animals on sick call.
4th	- Meet the assigned people and animals. - Get acquainted with the animals by talking, hand feeding, and petting them. **NOTE:** Hand feeding will teach the animal not to fear his handler and will associate the handler with food. As a reward for the animal, it is one of the best methods of gaining the animal's confidence. Hand feeding may replace the morning grain ration entirely or in part. - Learn to act in a firm, quiet manner.

Hours	Day 1 (Continued)

5th	• Demonstrate with a quiet animal the method of halter breaking with the aid of the haunch rope as follows:
	  ▪ Place a loop of rope over the hindquarters and run the free end through the halter. (This is the "haunch rope.")
	  ▪ Attach a lead rope to the halter ring to ensure control of the animal during training.
	  ▪ Keep the lead rope slack and call the animal while turning and walking away. If the animal does not follow, give a gentle tug on the rump rope without turning around.
	  ▪ Call the animal and give a command, such as "Come on," while pulling on the rope.
	  ▪ Tug on the lead rope to reinforce the call.
	NOTE: If the animal crowds the handler and clips his heels, the training should be stopped and the handler should get a lunging whip. He runs the whip through his hand and behind him. Then he repeats the leading exercise and, if crowded, waves the whip from side to side behind his back. He does not turn around while he is walking. The movement of the whip is usually enough to convince most animals to keep their distance. If all else fails, a smart tap on the nose or chest will usually settle matters. Some animals will try to walk beside the handler because of their initial halter training. This is not a problem on flat ground, but it can be a disaster in a narrow passage. If the animal seems disinclined to follow in the handler's footsteps, he should set up an obstacle course. The handler plants stakes or wire posts in the ground and strings cord between them to make a narrow track. He ties cloth along the lines so that the animal can see the markers.
	• Bring the animal up to the course and let it have a good look. (It may shy from the waving markers.)
	• Remain calm and speak soothingly while stroking the animal's head and neck; the animal will eventually follow the handler onto the track.
	• Make several passes so the lesson will sink in.
	NOTE: If the animal crashes through this barrier, it may be necessary to substitute cavallettis or jumps to create the course. When the handler stops and turns around, the animal should come up to him and stand quietly. The handler accomplishes this by giving a tug on the haunch rope and calling the animal's name. If this does not work, he tugs on the lead rope and calls again. When the animal is in front of him, he gives it a pat as a reward. The handler should discontinue the use of the haunch rope as soon as he is satisfied that the lessons have been learned.

6th	• Demonstrate halter breaking by use of the hackamore or a halter with a neck rope. • Continue until the animals respond to the rein aids before attempting to use bits. NOTE: A stubborn animal that will not lead should be snubbed to the saddle horn of the madrina or a mentor and pulled along. The handler walks in front of the animal with the halter shank attached to the halter and simulates leading the animal. A few days of this treatment will usually convince the animal that his handler is leading him. For control of a stubborn animal, the handler uses a neck rope placed around the neck with a half hitch around the nose. This restraint is also useful on animals that break away from their handlers.

Hours	Day 1 (Continued)
7th	• Play with the animals, brush them lightly with the brush, hand rub them, and do some currycomb operations. • Clean out their front feet and simulate shoeing by tapping the feet lightly with a stick or small stone.
8th	• Discuss the day's work and correction of faults. • Stress quietness and firmness in handling the animals and that the aim of the handlers is to eliminate fear of people in the animals and to gain their confidence.
Hours	**Day 2**
1st	• Catch the animals in the corral with food rewards after their name is called. • Exercise the animals lightly. • Walk the animals around the corral. • Ride the madrina while the other handlers form on the inside fence with ropes (similar to a racetrack). One person encourages the animals from the rear. • Stress quietness and safety.
2d	Tie up all the animals for a health inspection by the veterinarian.
3d	• Practice breaking and leading. • Use a haunch rope, if necessary.
4th	• Introduce the animals to "strange" things, such as ropes, saddle equipment, papers, raincoats, and blankets. • Drag a rope over the animals, around the feet, between the legs, across the head and neck, and over the back; let the rope fall over the haunches to the ground in the rear. (This must be done quietly.)
5th	• Mount all animals bareback. • Use the full hour. If the animals object, use more time. • Make no attempt to move the animals.
6th and 7th	Review 3d, 4th, and 5th hours.
8th	• Do the same as was done on the first day, including cleaning out the hind feet and simulating shoeing. • Discuss the day's work. • Note corrections and go over the next day's schedule.
Hours	**Day 3**
1st	Exercise the same as the second day.
2d	Tie up all the animals for a health inspection.
3d	Review the 3d, 4th, and 5th hours of the second day.
4th	• Introduce the animal to the blanket and roller. • Cinch lightly and walk the animals around.

Hours	Day 3 (Continued)
5th	• Introduce the saddle. • Lightly tighten the cinch, and lead the animals about with stirrup straps hanging down.
6th	• Tighten up the cinch and practice mounting and dismounting. • Use a mentor, if necessary.
7th	• Review the 3d, 4th, and 5th hours. • Include tying a raincoat loosely to the saddle. • Alternate tying to the pommel and cantle.
8th	• Increase hand rubbing, to include the pasterns, front and rear. • Simulate shoeing all around. • Discuss the day's work. • Note corrections and go over the next day's schedule.

Hours	Day 4
1st	Exercise the same as the second day.
2d	Tie up all the animals for a health inspection.
3d	Review the 3d, 4th, and 5th hours of the second day.
4th	Saddle up and practice mounting and dismounting on left and right sides.
5th	Practice starting and stopping, at a walk only.
6th	Review the 3d, 4th, and 5th hours of the second day.
7th	• Mount and turn right and left. • Mount and dismount.
8th	• Perform grooming tasks. • Discuss the day's work. • Note corrections and go over the next day's schedule.

Hours	Day 5
1st	Exercise the walk and trot alternately.
2d	Tie up all the animals for a health inspection.
3d and 4th	Review those phases of the previous training in which individual animals appear to be deficient.
5th	• Mount and drag a rope behind the animal while walking. • Practice placing ropes all over the animals.
6th	Practice starting, stopping, and turning right and left.
7th	Review all previous work and training.

Hours	Day 5 (Continued)
8th	<ul><li>Perform grooming tasks.</li><li>Discuss the day's work.</li><li>Apply corrections and go over the next day's schedule.</li></ul>

Hours	Day 6
1st	Exercise the same as the 5th day.
2d	Tie up all the animals for a health inspection.
3d, 4th, and 5th	Review all the previous work.
6th	<ul><li>Place a small load on the end of the rope for the animals to pull around.</li><li>Practice this exercise dismounted, and later mounted.</li><li>Use a load not to exceed 11 kilograms.</li><li>Make the animals pull straight away.</li></ul>
7th	<ul><li>Mount and practice starting, stopping, and turning right and left, at a walk only.</li><li>Practice picking up an object from the ground on the left and right sides while mounted.</li></ul>
8th	<ul><li>Perform grooming tasks.</li><li>Discuss the day's work.</li><li>Apply corrections and go over the next day's schedule.</li></ul>

Hours	Day 7
1st	<ul><li>Perform grooming and health inspection.</li><li>Rest the remainder of the day. (If proper care has been used during this first week, no animals will buck or run away. They will come when called. The animals should be in good health and on the way toward Body Condition Grade 3.)</li><li>Emphasize to all handlers the importance of being gentle and firm and the use of reward and punishment. (By now, all the animals should be quiet and easily caught.)</li><li>Commend the handlers who did outstanding work this week. (Some of the animals will have had some training prior to purchase. These animals will usually show more response to this schedule than green, unhandled animals. They should not be allowed to advance ahead of the slower animals.)</li><li>Keep all animals on the same schedule. It is better to go too slow than too fast.</li></ul>

Hours	Day 8
1st	<ul><li>Walk 30 minutes.</li><li>Walk and trot, alternately, 30 minutes.</li></ul>
2d	Tie up all the animals for a health inspection.
3d and 4th	Review all of the previous week's training.
5th	Mount and ride (walk only) all the remounts on a 5-kilometer march with the madrina at the head of the column.

Hours	Day 8 (Continued)
6th	Teach the animals to stand without being tied. NOTE: The objective is to be able to drop the line or reins and have the animal stand as if it were "ground tied." The rider dismounts dropping the reins and allowing them to dangle. The reins must not be tied. The animal will try to follow due to its previous training of the haunch rope. A lair rope is used to tie the animals to a stake in the ground. It is one centimeter thick and ten meters long with an eye sliced at one end. The other end must be wrapped or seized. Once the animal shows signs of staying put, the line only is dropped on the ground. If this works, the reins only are dropped and the animal is tested to see if it will stay put. This action is repeated as long as necessary.
7th	Mount and ride (walk only) the animals double.
8th	• Perform grooming tasks. • Discuss the day's work. • Apply corrections and go over the next day's schedule.

Hours	Day 9
1st	Exercise the same as the 8th day.
2d	Tie up all the animals for a health inspection.
3d and 4th	Review the previous week's training.
5th	Conduct a cross-country ride (walk and trot) of 5 km.
6th	Teach the animals to stand without being tied. (See Note on Day 8, 6th Hour.)
7th	Lead the animals from the right side.
8th	• Perform grooming tasks. • Discuss the day's work. • Apply corrections and go over the next day's work.

Hours	Day 10
1st	Exercise the same as the 8th day.
2d	Tie up all the animals for a health inspection.
3d and 4th	Conduct an 8-km march with the bell mare at the head of the column.
5th	Mount and ride (walk only) all the remounts on a 5-km march with the madrina at the head of the column.
6th	Teach the animals to stand without being tied. (See Note on Day 8, 6th Hour.)
7th and 8th	• Mount and ride (walk only) all the remounts on a 5-km march with the madrina at the head of the column. • Repeat training to ground tie. • Use lair rope, if needed.

Hours	Day 11
1st	Tie up all the animals for a health inspection.
2d and 3d	Conduct an 8-km cross-country march as individual riders.
4th	• Review the 6th hour of the 8th day. • Simulate saddling with a packsaddle. Two handlers place riding saddles on each animal from the rear.
5th	• Have the animals pull a 50-kg weight for 30 minutes by a rope. • Practice for 30 minutes with mounted mules, using a can filled with rattling objects (lots of noise).
6th	• Practice for 30 minutes carrying a load on the saddle while riding double. • Lead from the right side for 30 minutes.
7th	• Introduce the animals to the packsaddle. • Saddle up and adjust cinches and breeching (brichen). • Lead the animals around the corral.
8th	• Perform grooming tasks. • Discuss the day's work. • Apply corrections and go over the next day's schedule.

Hours	Day 12
1st	Tie up all the animals for a health inspection.
2d and 3d	Saddle the animal with a packsaddle for a 5-km march (saddle only) with the madrina at the head of the column.
4th and 5th	Review all of the first week's training.
6th and 7th	Review all of this week's work.
8th	• Perform grooming tasks. • Discuss the day's work. • Apply corrections and go over the next day's schedule.

Hours	Day 13
1st	Tie up all the animals for a health inspection.
2d and 3d	• Saddle the animal with a packsaddle only for an 8-km march with the madrina at the head of the column. • Lead from the left and right sides while dismounted.
4th	Review the 6th hour of the 8th day.
5th	Review the 7th hour of the 10th day.
6th	Review the 5th hour of the 11th day.
7th	• Introduce the animals to "strange" things; lead them over obstacles, such as improvised bridges, ditches, and so on. • Do not allow the animals to jump; hold their heads down.

Hours	Day 14
1st through 8th	• Tie up all the animals for a health inspection. • Rest. NOTE: The animals should be in good health and on the way toward better condition. The handlers should keep the animals coming along slowly, but also should keep increasing the amount of work. The goal is Body Condition Grade 3 and the ability to march twenty miles a day under full payloads of 115 kg, plus packsaddle. Daily marches under full payloads must be maintained to keep in Grade 3. If the animals go backward in condition through lack of work and then are marched 35 to 45 km under full payloads in hot weather, they will die.
Hours	**Day 15**
1st	Tie up all the animals for a health inspection.
2d	• Select the top-load, side-load, and saddle animals. NOTE: The handler should pay particular attention to selecting animals for top loads, and consider conformation, such as straight backs and low croup. • Identify gentle, steady, and smooth-gaited animals for medical evacuation duties.
3d and 4th	• Saddle all pack animals for a 5-km march. • Rest the riding animals during this period.
5th, 6th, and 7th	• Saddle the riding animals for a 5-km individual march. • Walk and trot alternately. • Practice, while having the animals ground tied, mounting from left and right sides, and pulling a 50-kg weight. • Ride double. NOTE: The pack masters and cargadores start fitting packsaddles to the pack animals during this period.
8th	• Perform grooming tasks. • Discuss the day's work. • Apply corrections and go over the next day's schedule.
Hours	**Day 16**
1st through 8th	• Tie up all the animals for a health inspection. • Rest. • Take the animals on a pleasure ride, if possible. NOTE: A picnic at midday is a good treat for everyone.
Hours	**Day 17**
1st	Tie up all the animals for a health inspection.
2d, 3d, and 4th	• Saddle the pack animals with 20-kg single loads for an 8-km march over obstacles and slightly difficult terrain. • Rest the riding animals.

Hours	Day 17 (Continued)
5th, 6th, and 7th	• Saddle the riding animals for an 8-km march. NOTE: The pack masters and cargadores continue fitting saddles to the pack animals.
8th	• Perform grooming tasks. • Discuss the day's work. • Apply corrections and go over the next day's schedule.
Hours	**Day 18**
1st	Tie up the animals for a health inspection.
2d, 3d, and 4th	Saddle the pack animals with 35-kg single loads for an 8-km march.
5th, 6th, and 7th	Saddle the riding animals for an 11-km march over obstacles and difficult terrain.
8th Hour	• Perform grooming tasks. • Discuss the day's work. • Apply corrections and go over the next day's schedule.
Hours	**Day 19**
1st	Tie up all the animals for a health inspection.
2d, 3d, and 4th	Saddle pack animals with 50-kg single loads for an 8-km march (same as the 17th day).
5th, 6th, and 7th	Same as the 18th day, except will be a 13-km march.
8th	• Perform grooming tasks. • Discuss the day's work. • Apply corrections and go over the next day's schedule.
Hours	**Day 20**
1st	Tie up all the animals for a health inspection.
2d, 3d, and 4th	Same as the 19th day, except will be a 13-km march.
5th, 6th, and 7th	Same as the 18th day, except will be a 16-km march.
8th	Perform grooming tasks.
Hours	**Day 21**
1st	Tie up all the animals for a health inspection.
2d	Prepare for a competitive inspection.
3d, 4th, and 5th	Conduct a competitive inspection.

Hours	Day 21 (Continued)
6th	- Perform grooming tasks. - Turn animals out to graze and rest.
7th and Beyond	- Give awards for competitive inspection and a party for the entire staff. - Release animals from quarantine if the veterinarian orders no further quarantine. All animals are released for duty. - Single out the good handlers and commend them accordingly. - Make sure the commendation is noted in their records.

Appendix C

Animal Packing Formations

This appendix provides illustrated formations that a pack animal detachment can use to conduct special operations missions in different environments around the world. These formations will vary based on the detachment's size, terrain, enemy characteristics, and the overall mission objective.

A-1. There are three types of formations included in this appendix. Within each type are various examples that show how changes must occur to compensate for different scenarios.

A-2. The types of formations are as follows:

- Basic Animal Packing:
  - Column Formations V-1 and V-2, Figure C-1, page C-2.
  - Traveling Overwatch Formations, Figure C-2, page C-3.
- Danger Areas:
  - Linear Danger Area, Figure C-3, pages C-4 through C-6.
  - Large Open Area Extended Overwatch, Figure C-4, pages C-7 through C-9.
  - Large Open Area Traveling Overwatch, Figure C-5, pages C-10 and C-11.
  - Shallow River Crossing, Figure C-6, pages C-12 through C-14.
- Immediate Action Drills:
  - Hasty Ambush Defensive Measure V-1, Figure C-7, pages C-15 through C-17.
  - Hasty Ambush Defensive Measure V-2, Figure C-8, page C-18.
  - Hasty Ambush Offensive Measure, Figure C-9, pages C-19 and C-20.
  - Air Attack, Figure C-10, pages C-21 and C-22.

FM 3-05.213

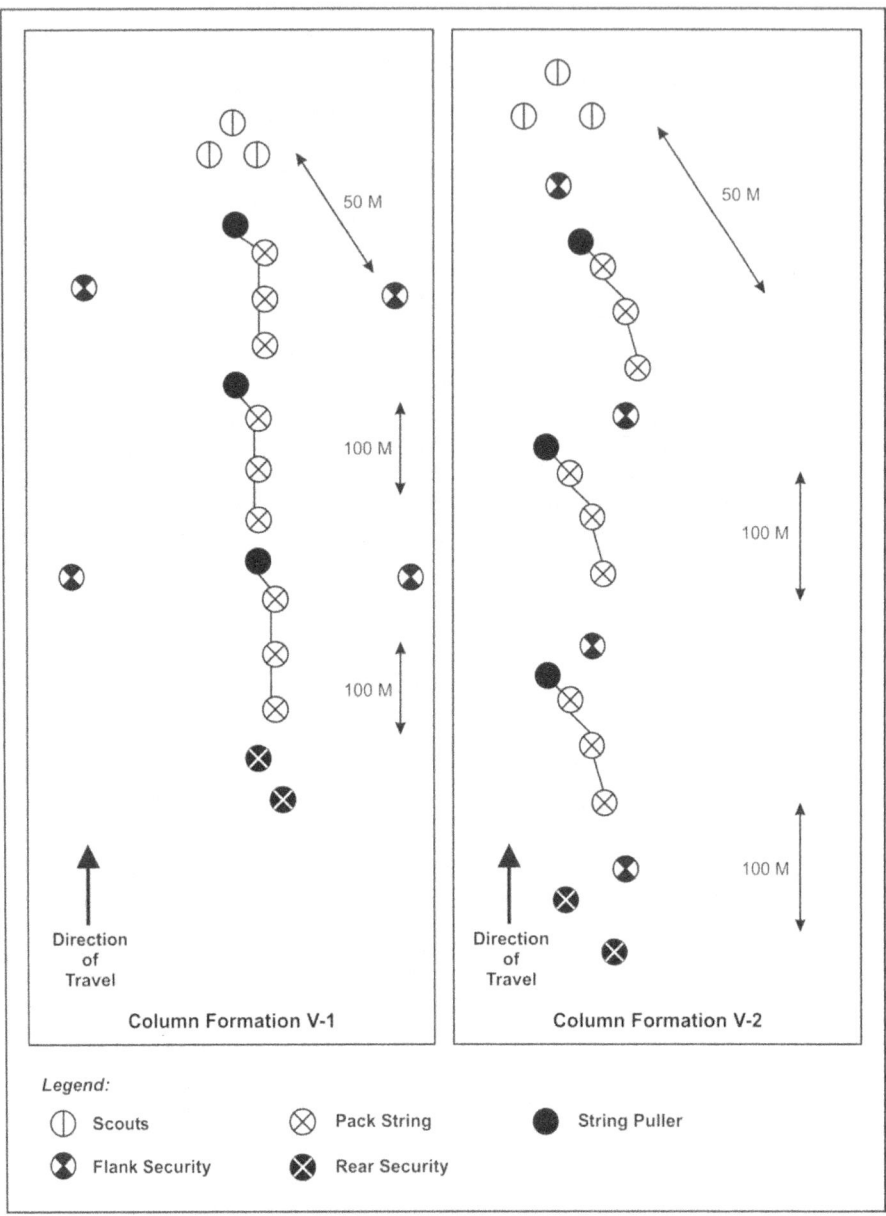

Figure C-1. Column Formations

C-2

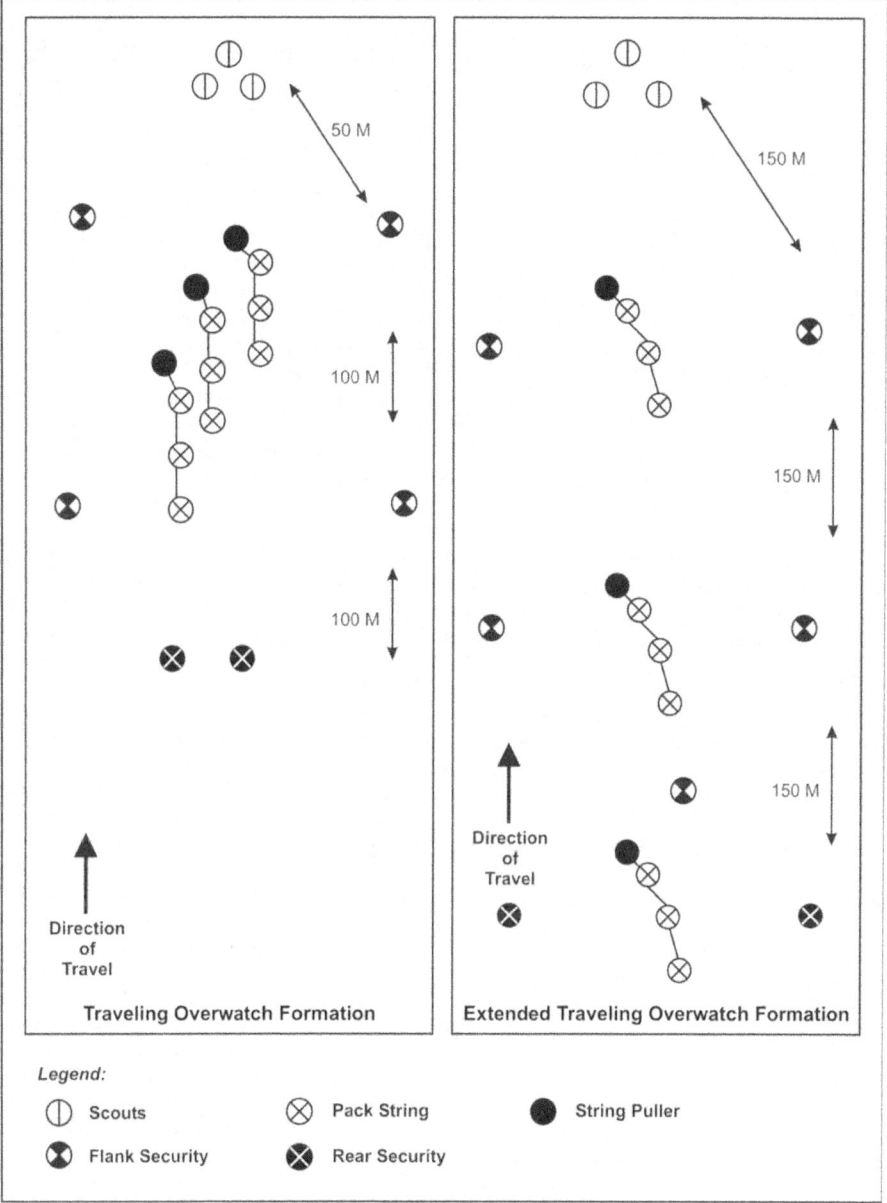

Figure C-2. Traveling Overwatch Formations

FM 3-05.213

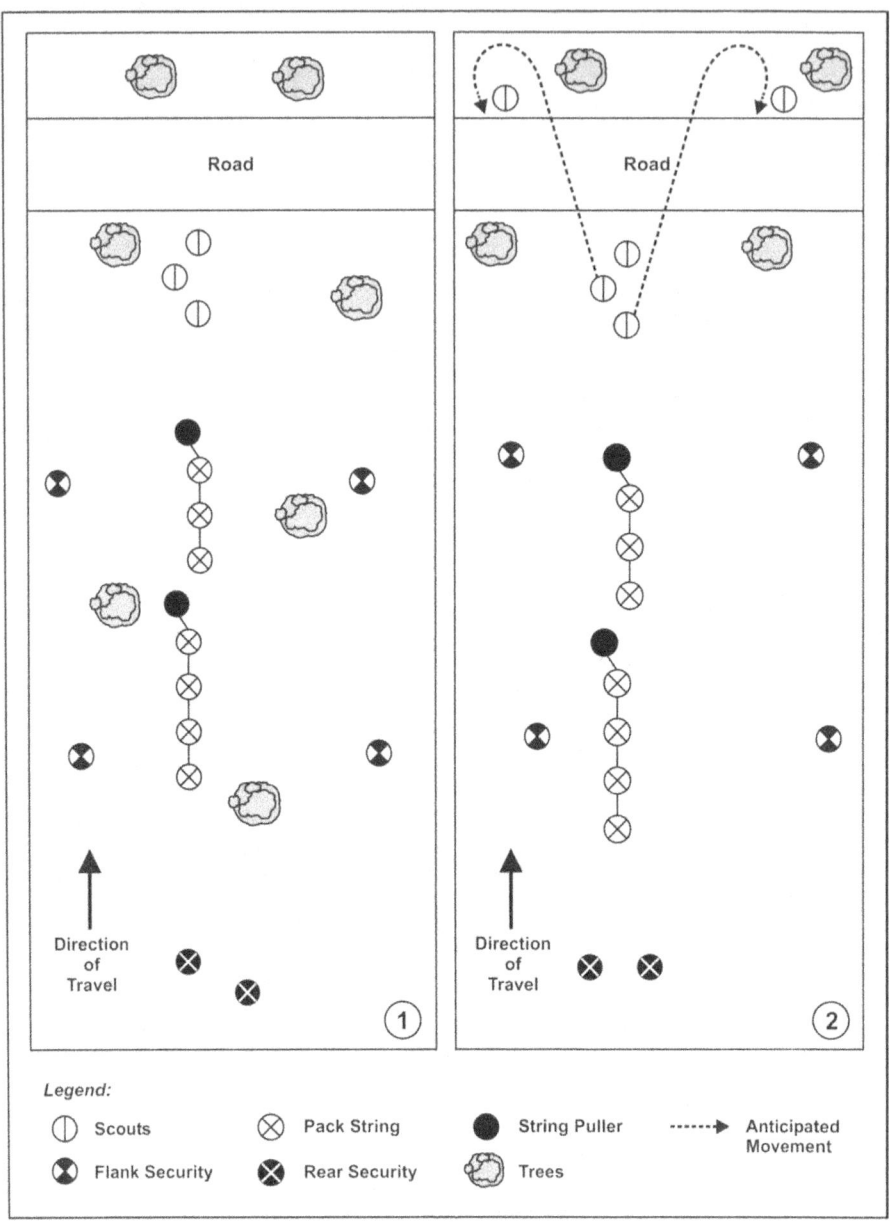

Figure C-3. Linear Danger Area

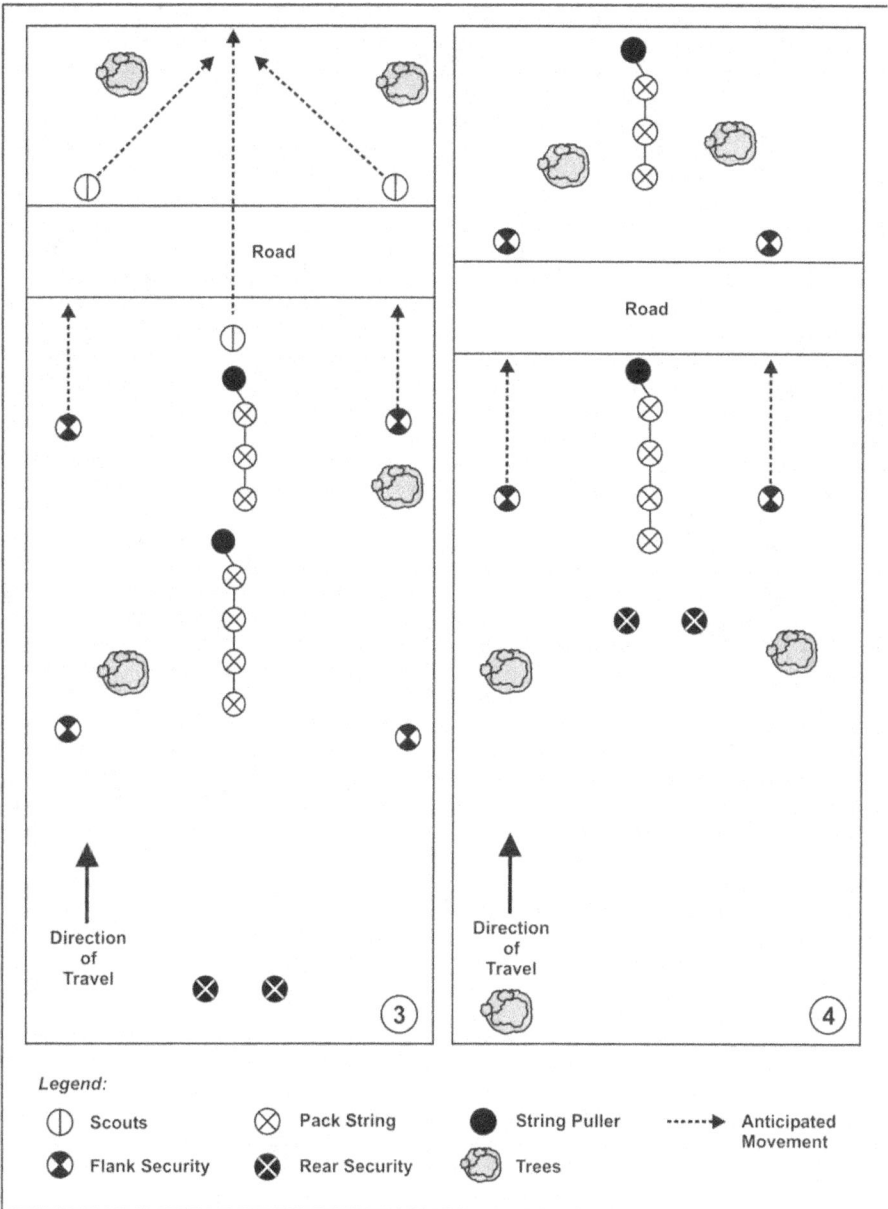

Figure C-3. Linear Danger Area (Continued)

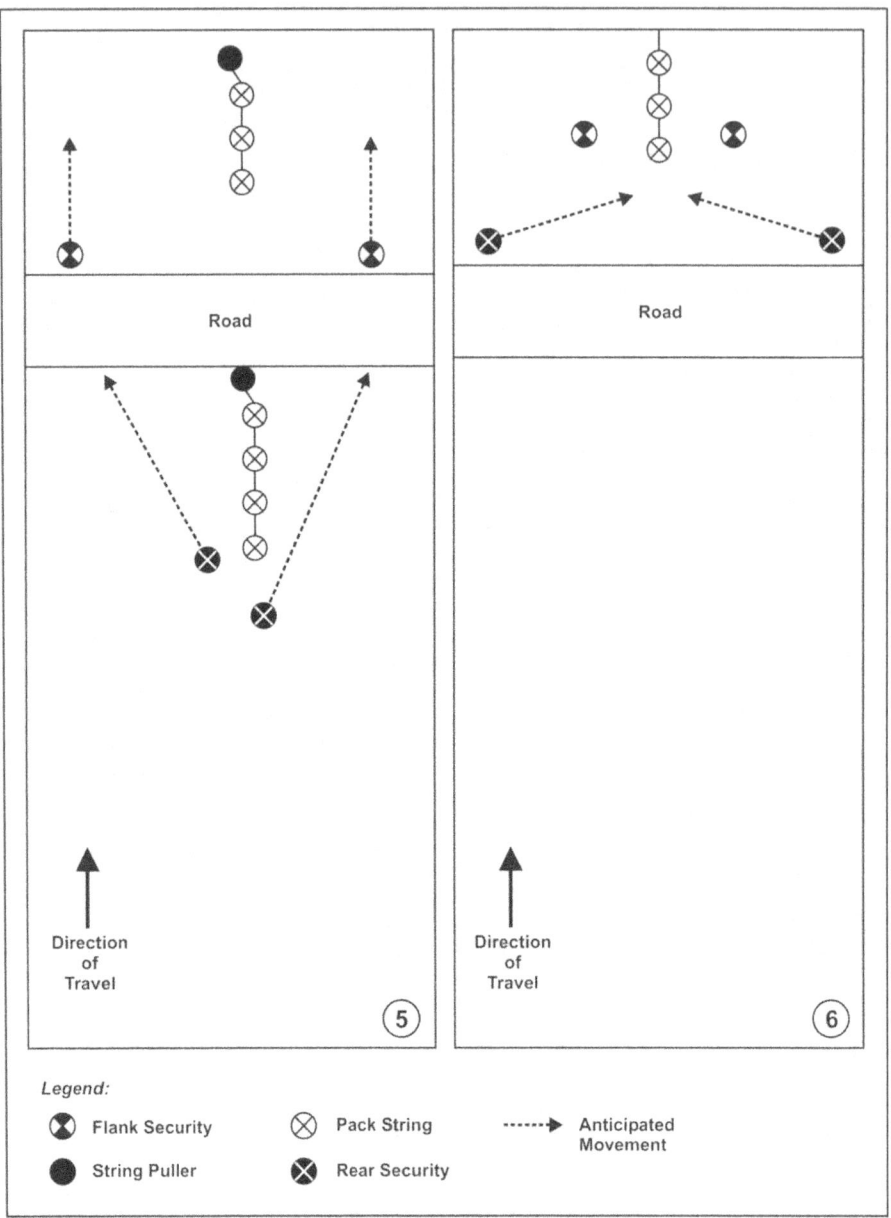

Figure C-3. Linear Danger Area (Continued)

FM 3-05.213

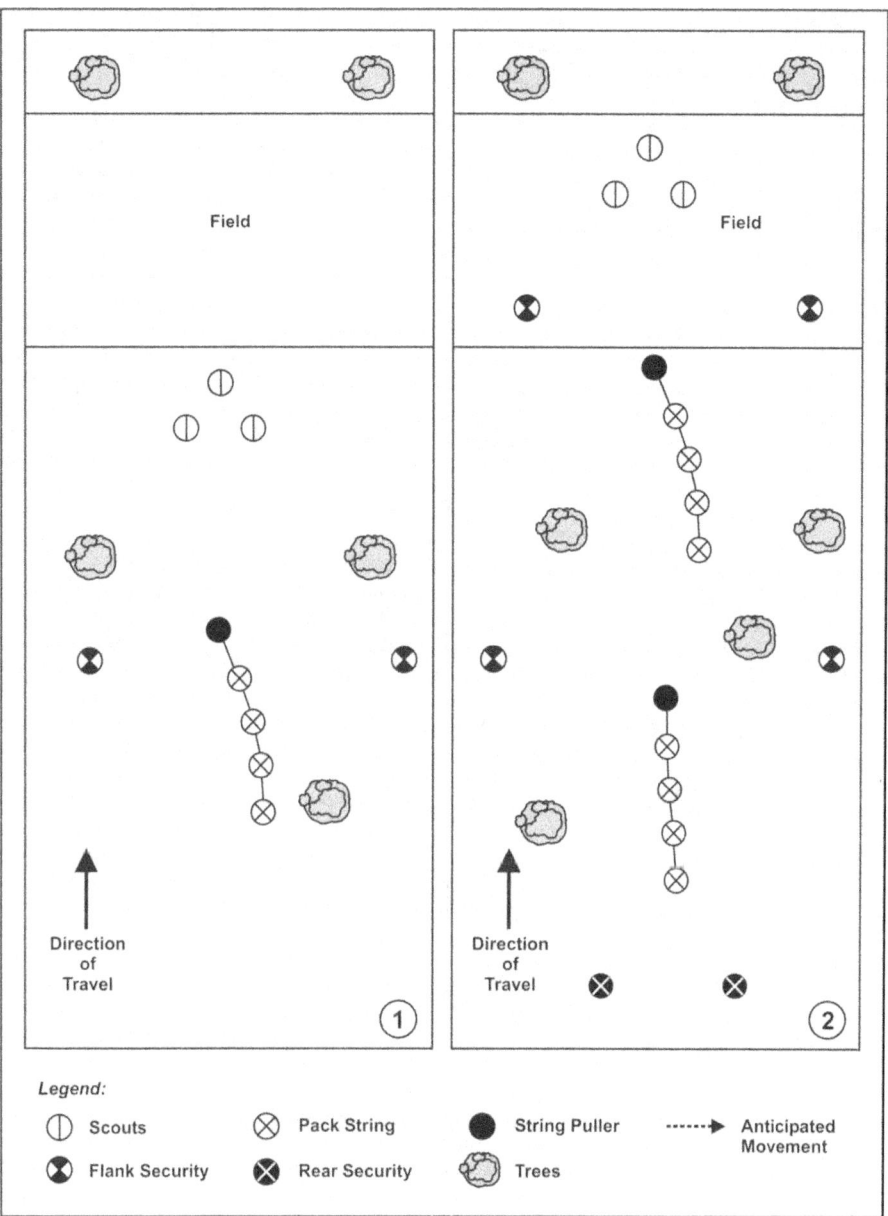

Figure C-4. Large Open Area Extended Overwatch

C-7

FM 3-05.213

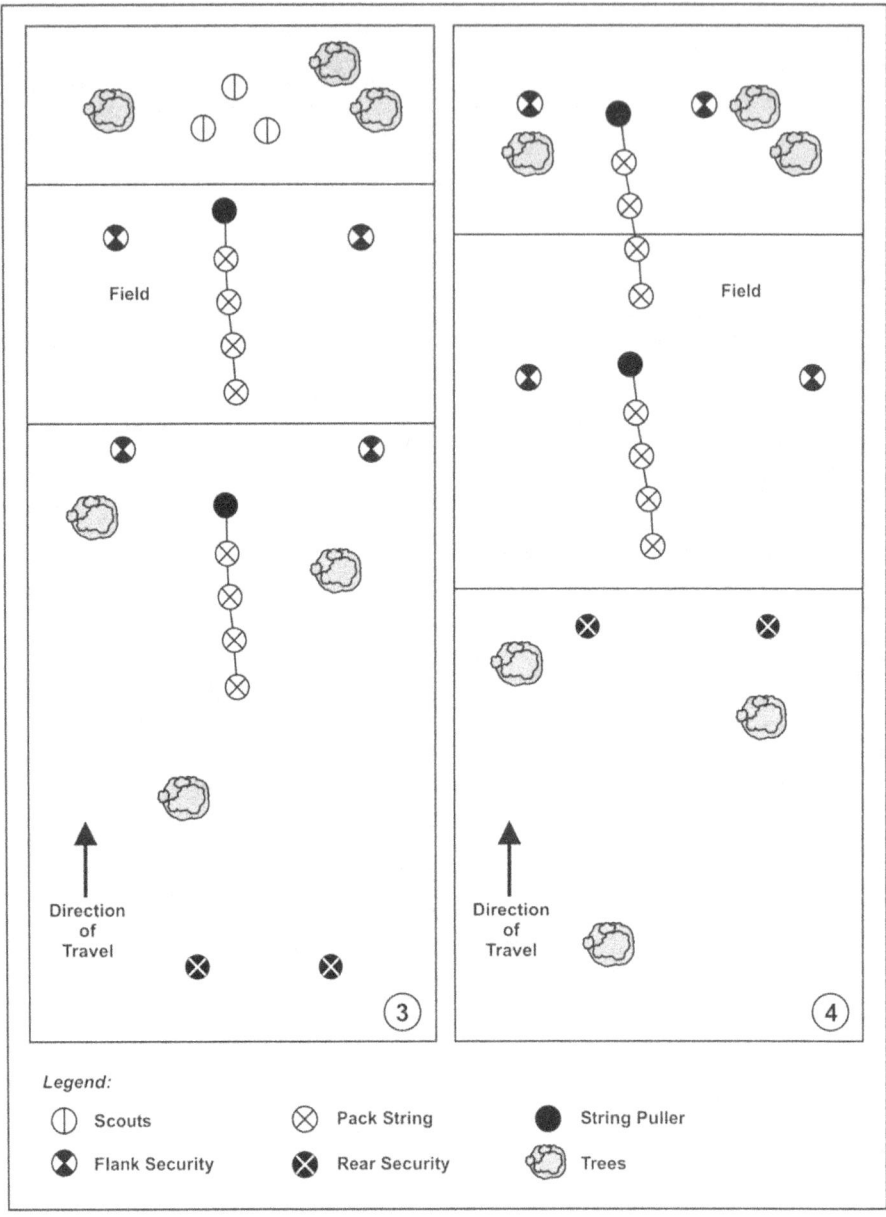

Figure C-4. Large Open Area Extended Overwatch (Continued)

C-8

FM 3-05.213

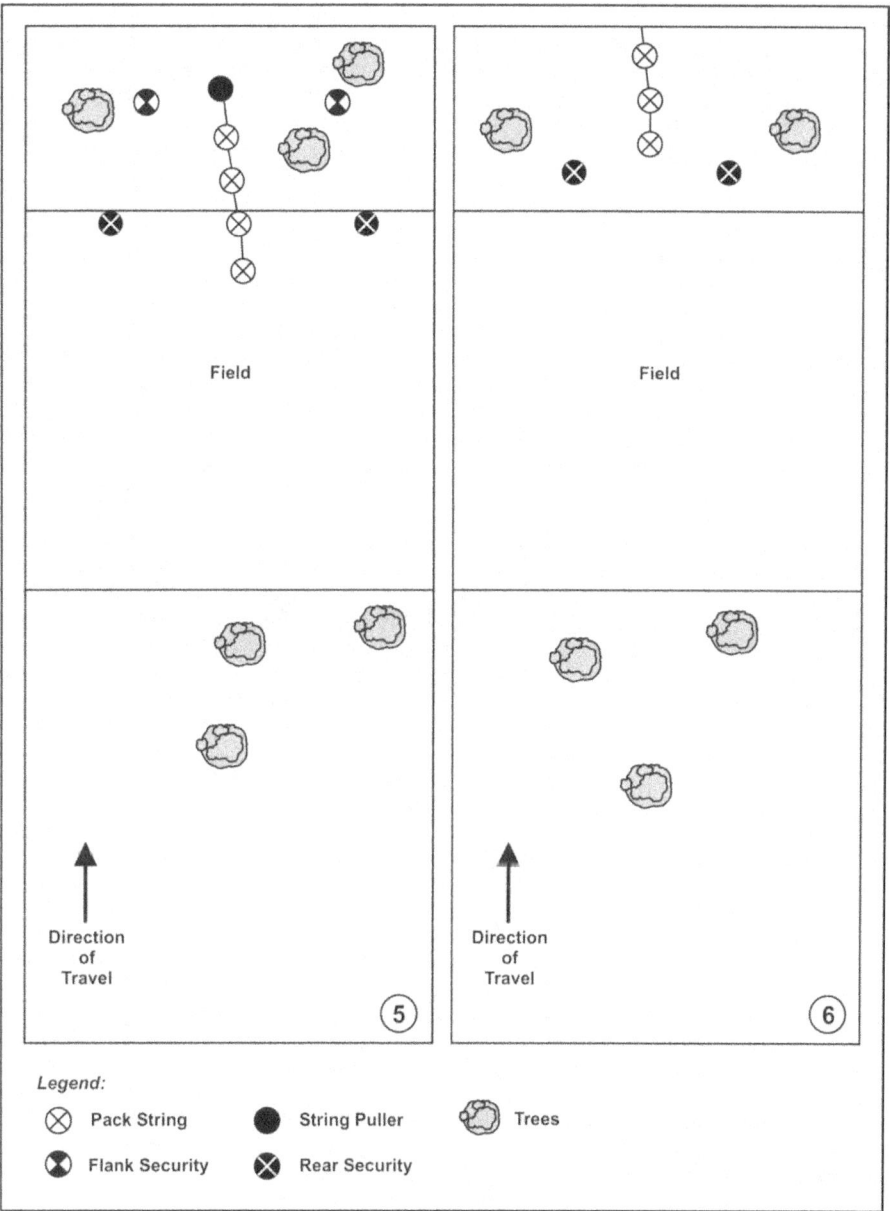

Figure C-4. Large Open Area Extended Overwatch (Continued)

C-9

FM 3-05.213

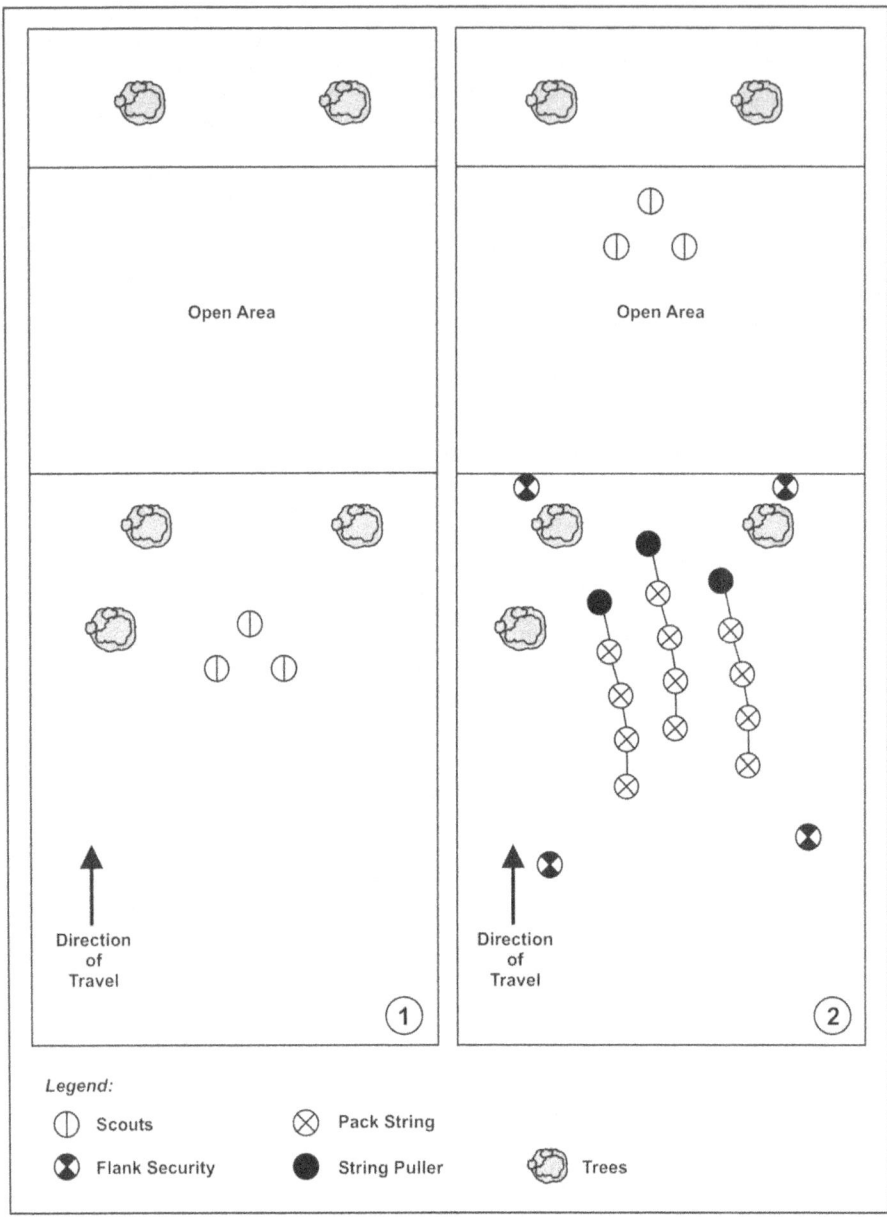

Figure C-5. Large Open Area Traveling Overwatch

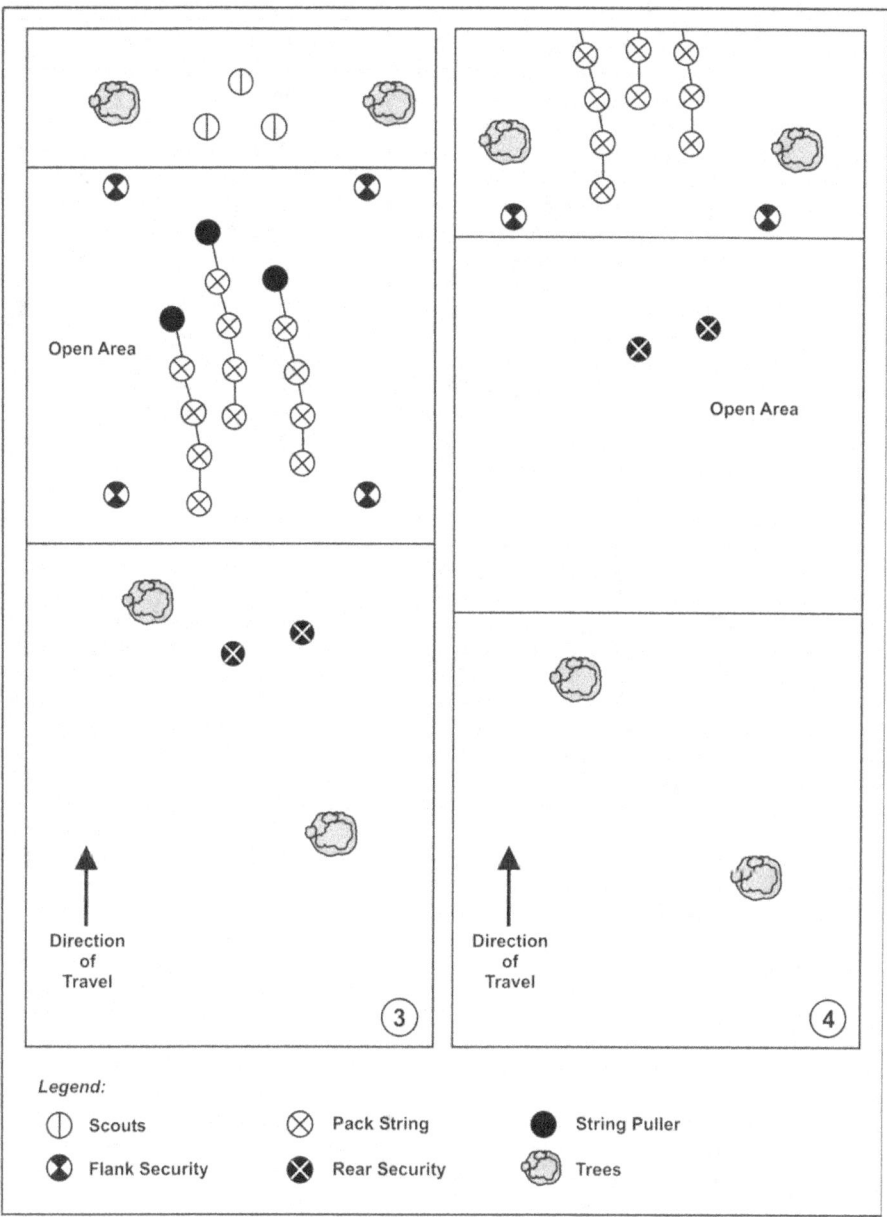

Figure C-5. Large Open Area Traveling Overwatch (Continued)

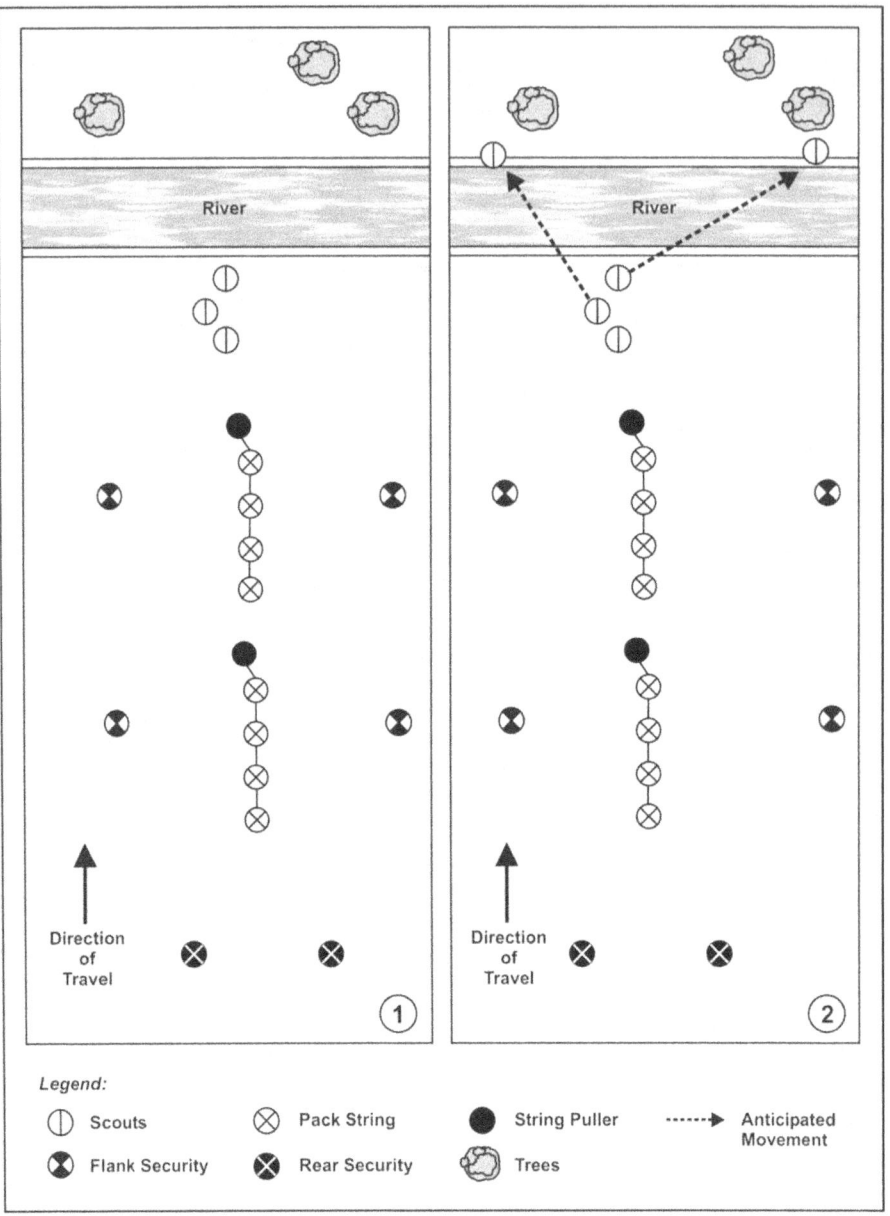

Figure C-6. Shallow River Crossing

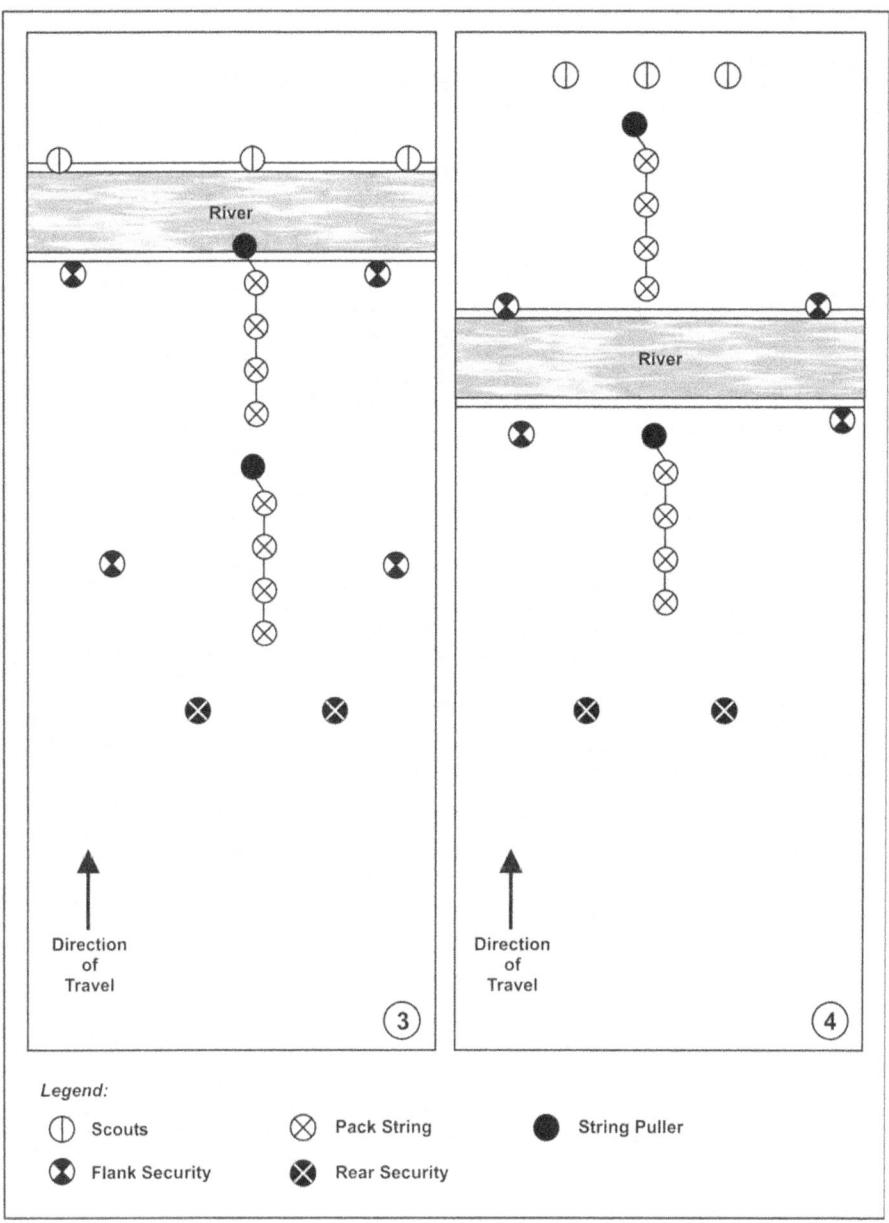

Figure C-6. Shallow River Crossing (Continued)

FM 3-05.213

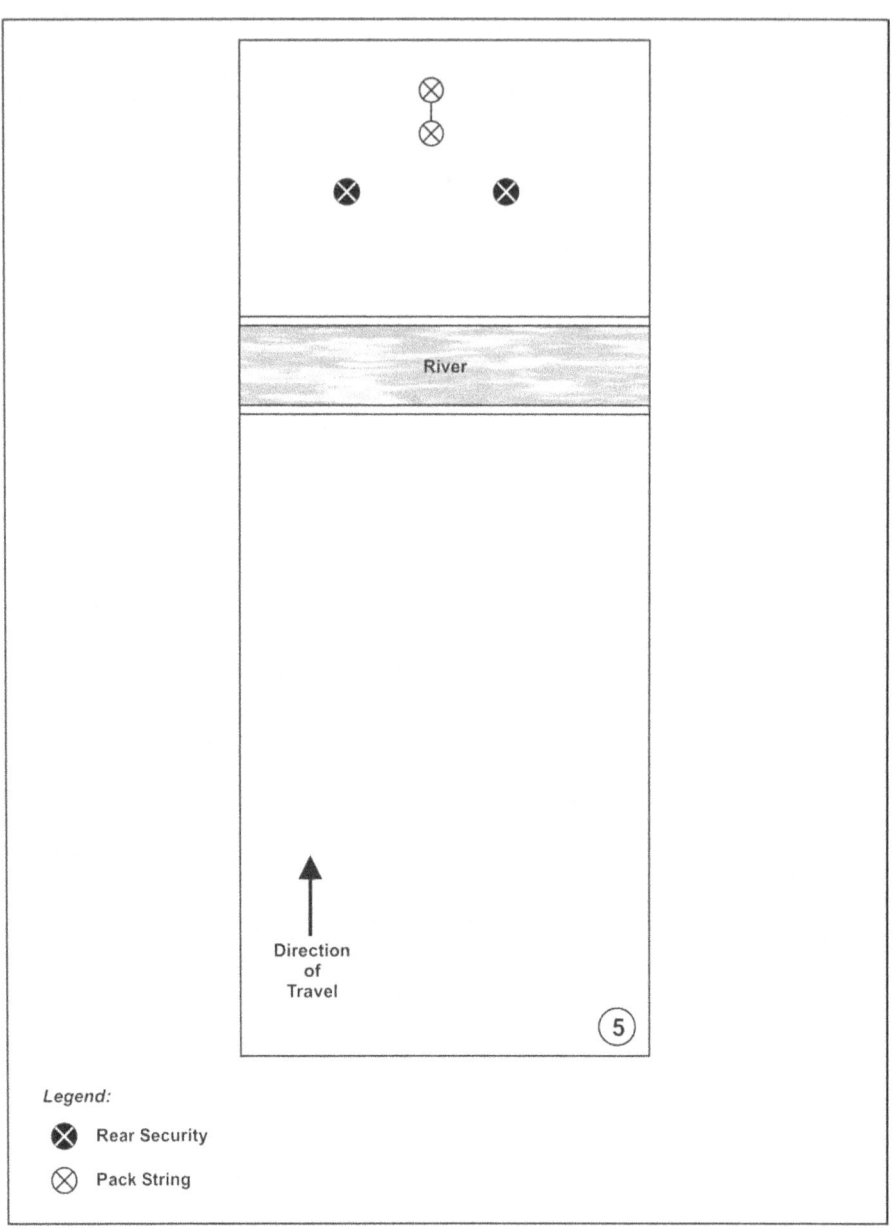

Figure C-6. Shallow River Crossing (Continued)

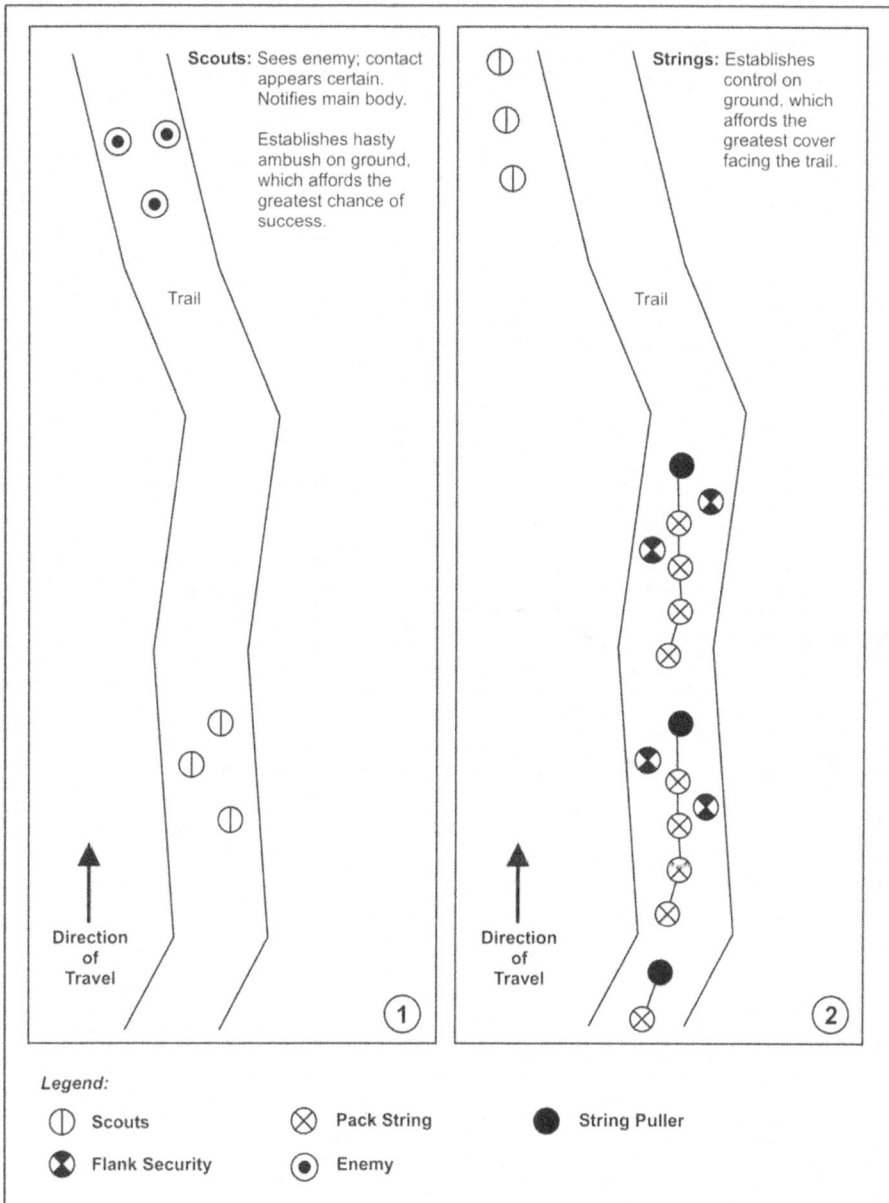

Figure C-7. Hasty Ambush Defensive Measure V-1

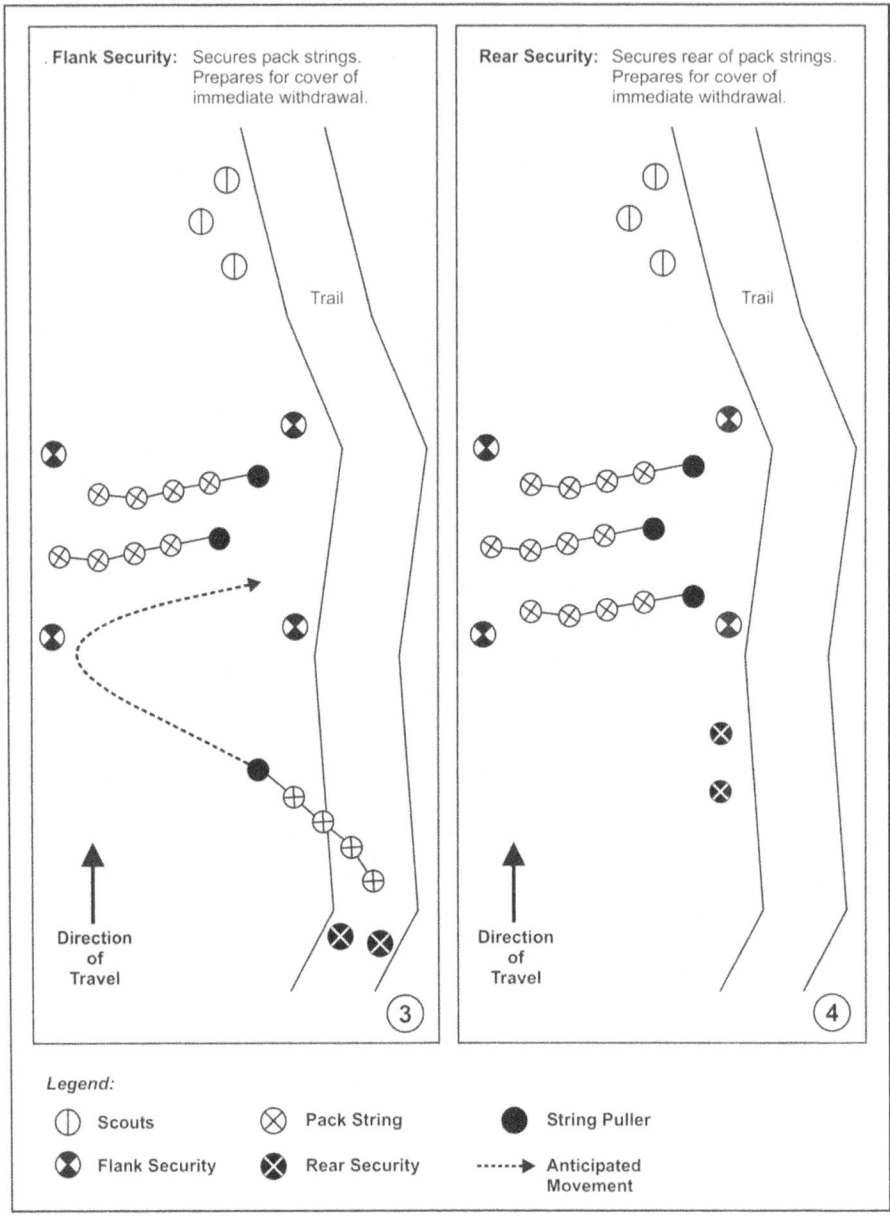

Figure C-7. Hasty Ambush Defensive Measure V-1 (Continued)

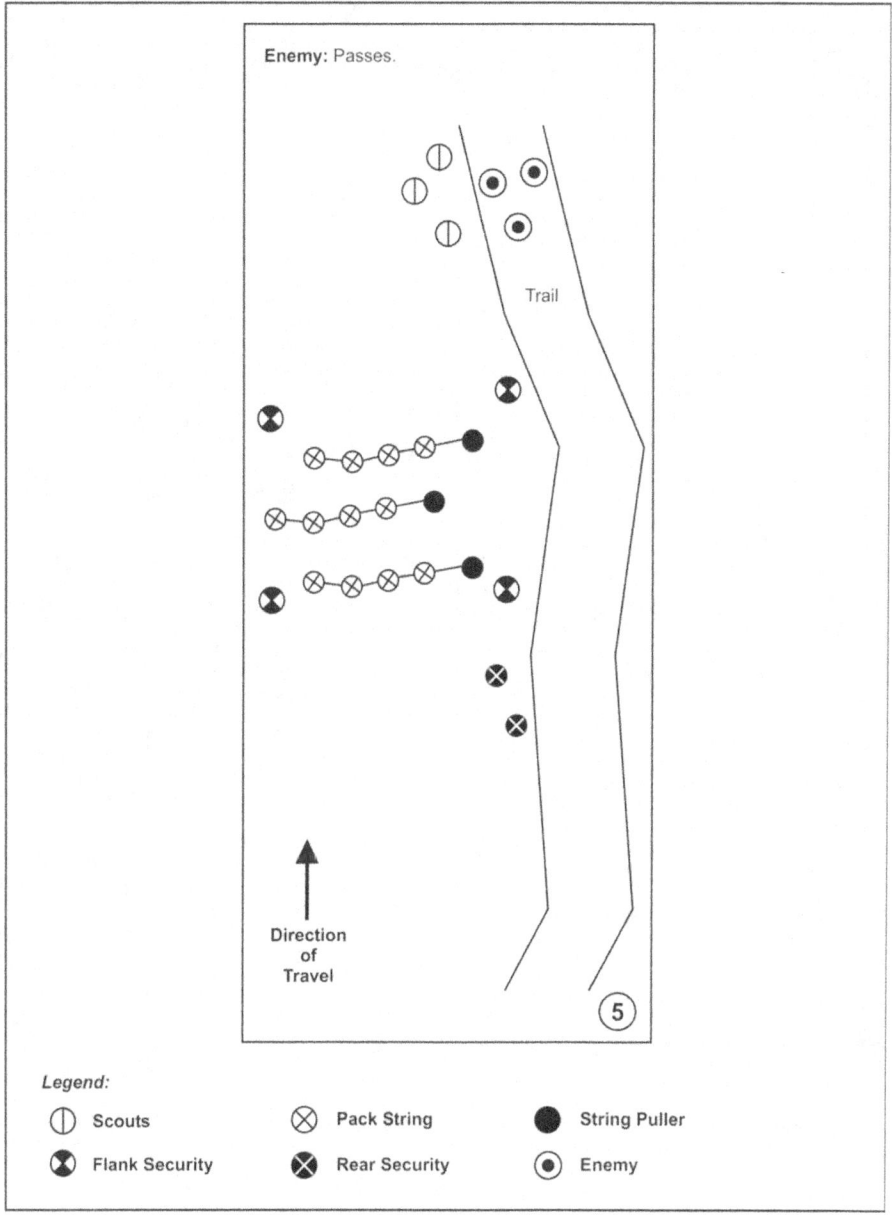

Figure C-7. Hasty Ambush Defensive Measure V-1 (Continued)

FM 3-05.213

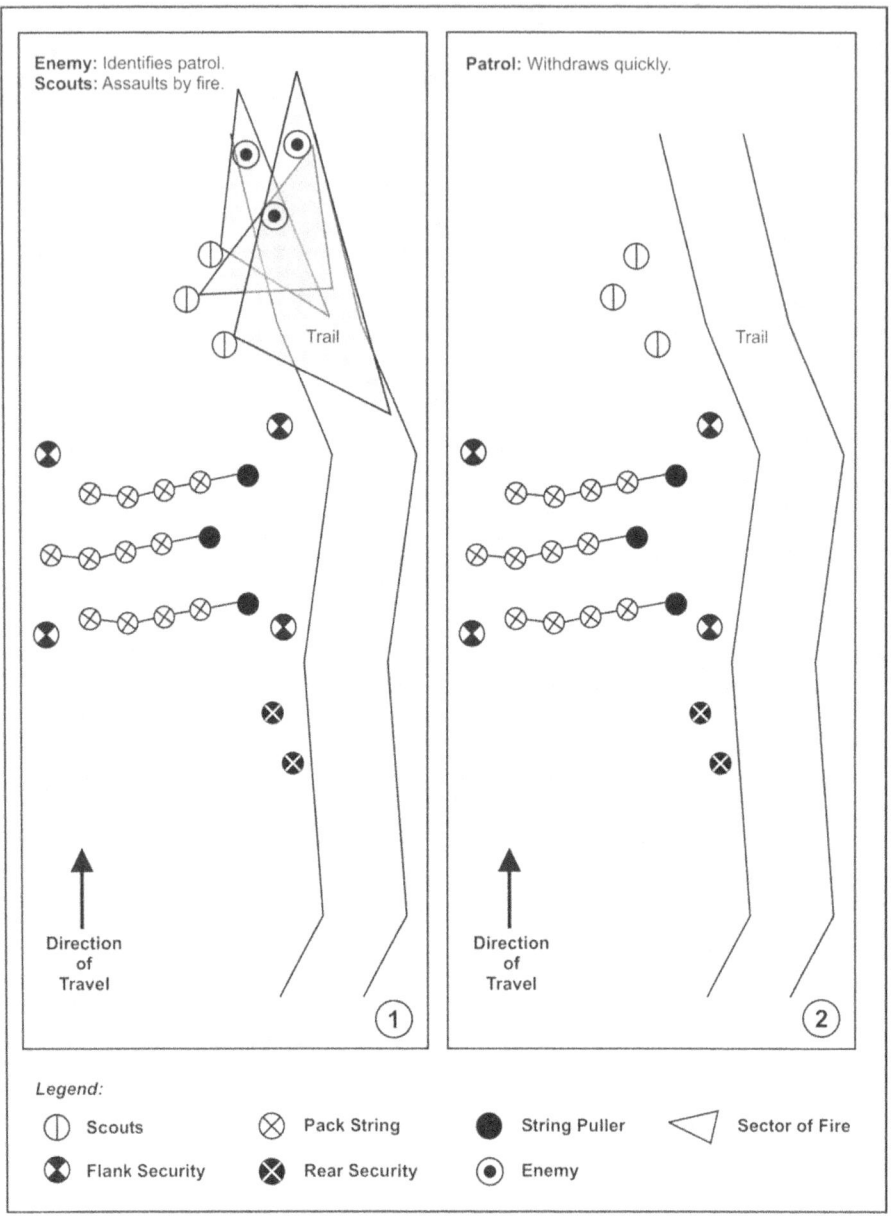

Figure C-8. Hasty Ambush Defensive Measure V-2

FM 3-05.213

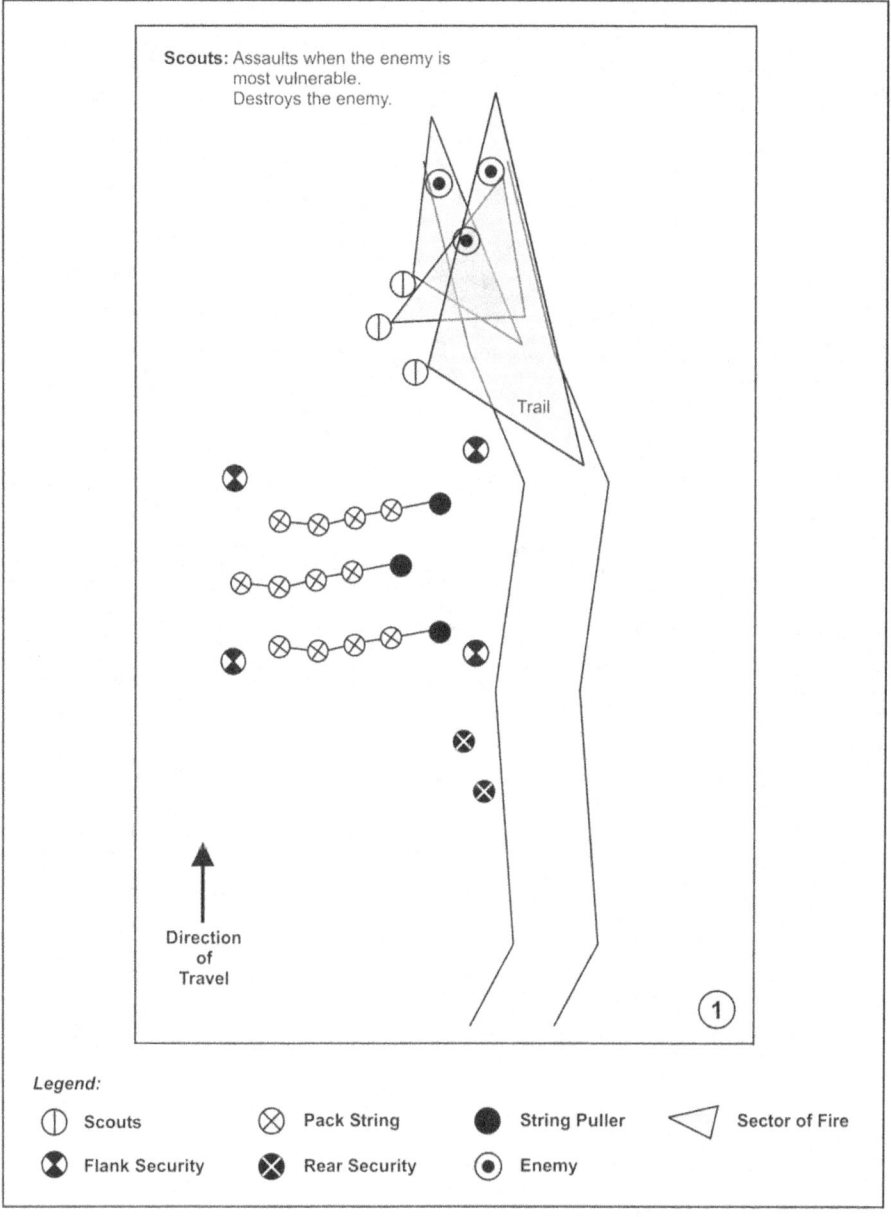

Figure C-9. Hasty Ambush Offensive Measure

C-19

FM 3-05.213

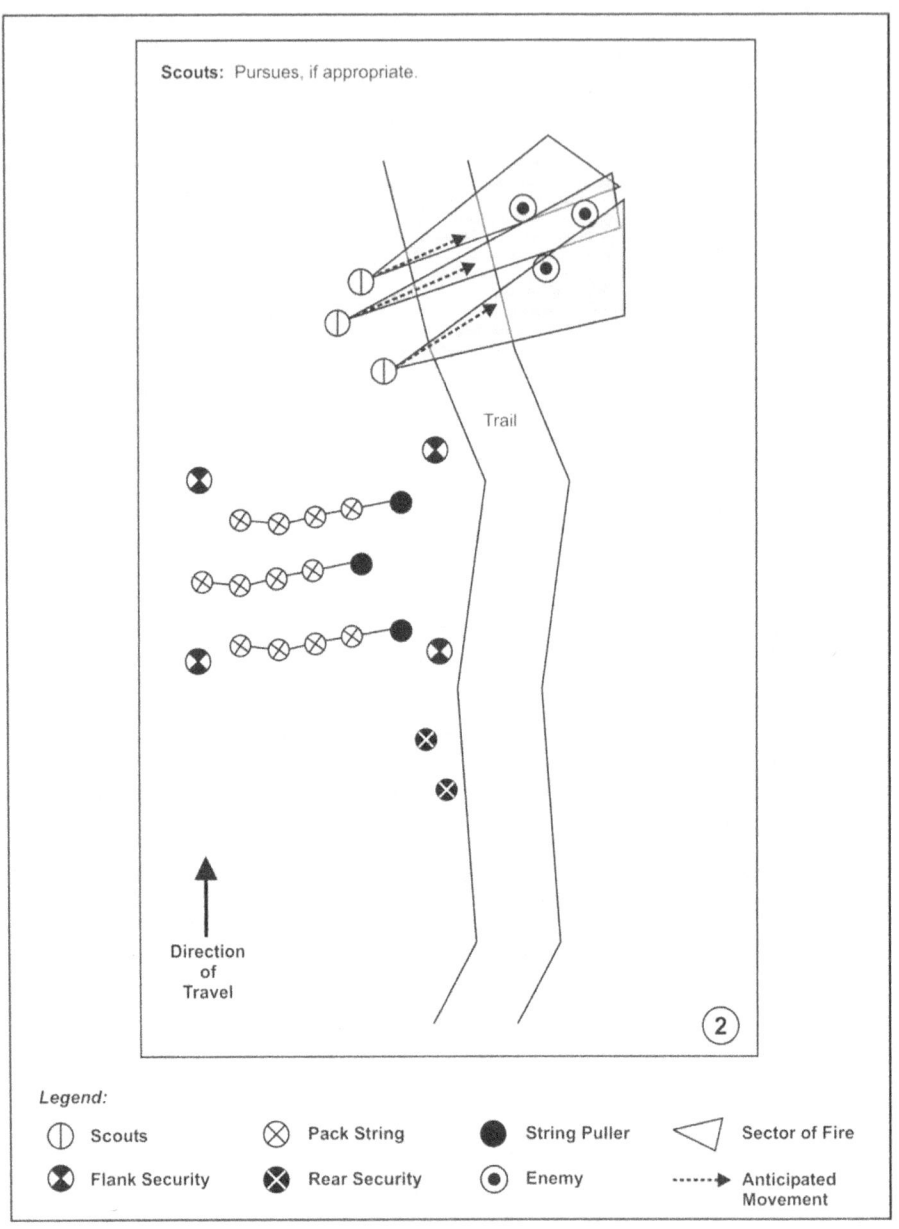

Figure C-9. Hasty Ambush Offensive Measure (Continued)

FM 3-05.213

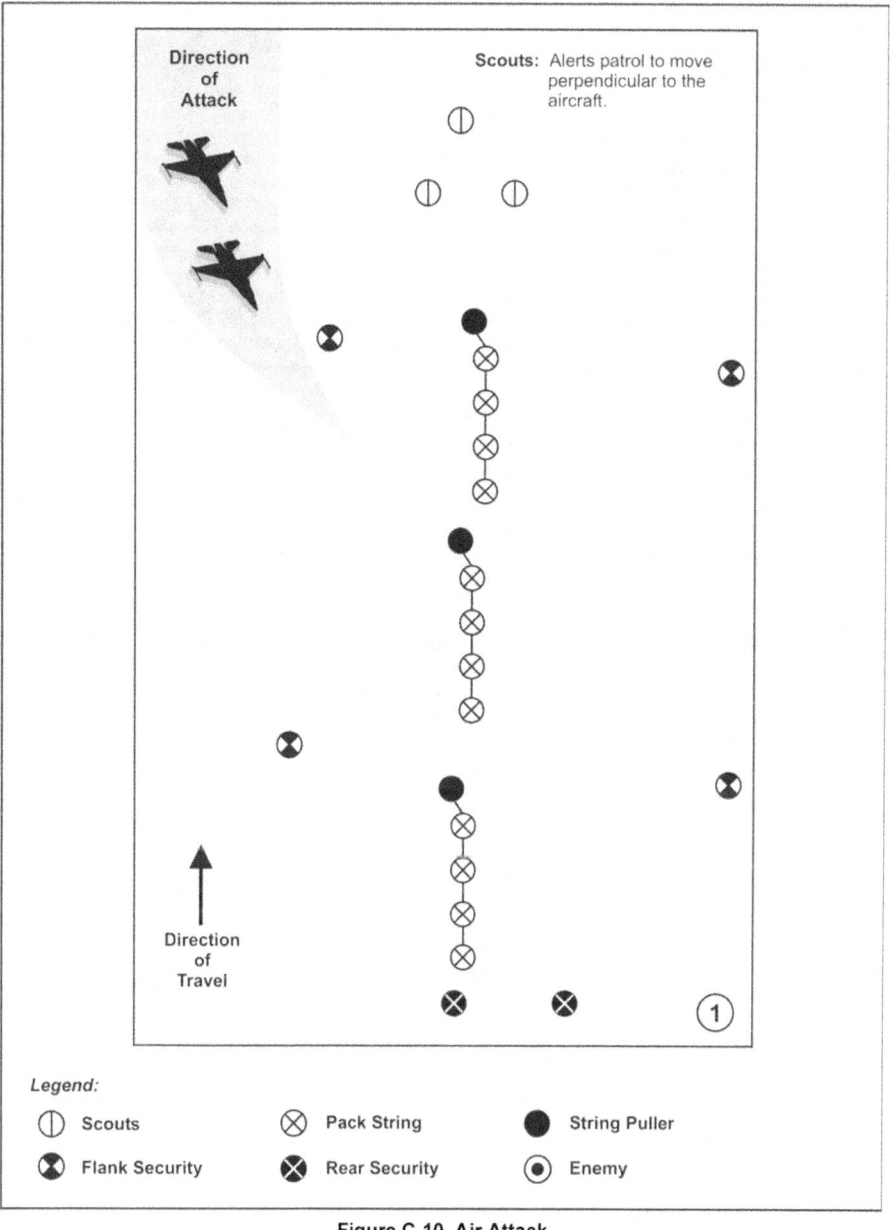

Figure C-10. Air Attack

C-21

FM 3-05.213

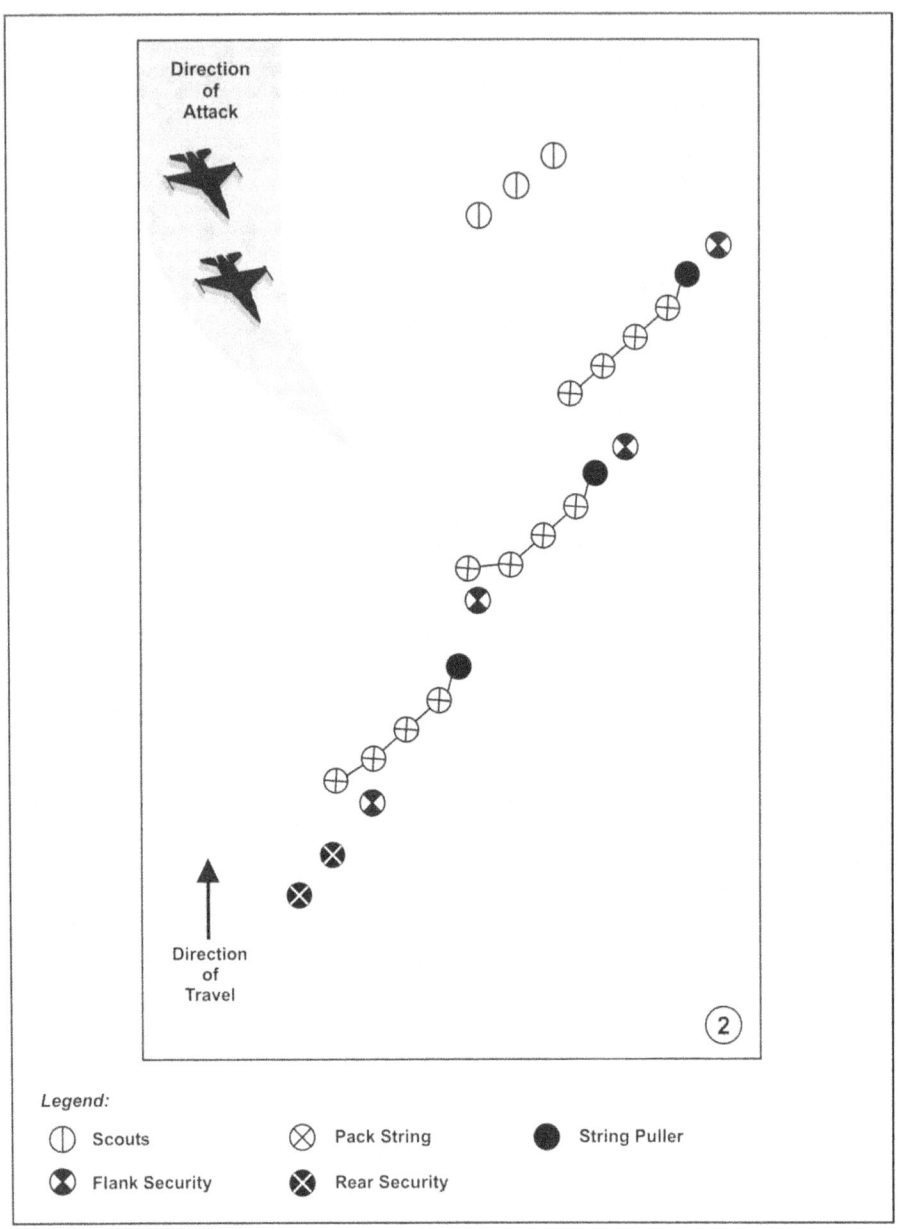

Figure C-10. Air Attack (Continued)

C-22

Glossary

ALICE	all-purpose, lightweight, individual carrying equipment
AO	area of operations
ARSOF	Army special operations forces
ass/burro/donkey	Proper interchangeable terms that mean the same thing.
auger	Any of various tools or devices used for boring holes or moving loose material.
BCS	body condition score
Bell Mare	The lead horse.
bpm	beats per minute
btl	bottle
C	Celsius
cargadore	A native of the operational area or a detachment member that handles the animals and may be used as a guide.
cm	centimeter(s)
CNS	central nervous system
CPR	cardiopulmonary resuscitation
cross tie	A mild form of restraint in which the animal's head is secured in a normal raised position by two tie ropes extending from the ring in the halter to opposite sides of the stall or between two trees.
ea	each
ear-ial	Telegraphing of a donkey's voice.
easy boot	Polyurethane boot to provide hoof protection.
EEE	Eastern equine encephalomyelitis
EIA	equine infectious anemia
F	Fahrenheit
farrier	A person who shoes horses.
FM	field manual
foal	A baby donkey
forb	A broad-leaved herbaceous plant, as distinguished from the grasses, sedges, shrubs, and trees.
g	gram(s)
gal	gallon(s)

gelding	A castrated male donkey.
general support	Support that is given to the supported force as a whole and not to any particular subdivision thereof. (FM 101-5-1)
GRAIL	Soviet SA-7 surface-to-air missile
hackamore	A bridle with a loop capable of being tightened about the nose in place of a bit or with a slip noose passed over the lower jaw.
halter	A device used to control an animal.
hand	A measuring unit equal to 4 inches.
harborage	A place of shelter or lodging.
HE	high explosive
HEENT	head, ears, eyes, nose, and throat
hinny	The hybrid animal produced when the female ass (jennet) is mated to the male horse (stallion) to produce a foal.
horse mule	A male mule.
hr	hour(s)
I&D	incise and drain
IM	intramuscular
ISBN	International Standard Book Number
IU	immunizing unit
IV	intravenous
jack	A male donkey, mostly refers to intact males.
jennet/jenny	A female donkey.
john	A male mule.
kg	kilogram(s)
km	kilometer(s)
LAW	light antitank weapon
lb	pound(s)
LCE	load-carrying equipment
loads	Equipment or supplies that are packed into a mantee or pannier to be transported by the pack animal.
madrina	An animal (usually an old mare) wearing a bell and acting as the leader of a pack string.
mantee	A piece of canvas, from 7 by 7 feet up to 10 by 12 feet, used to throw over a load or to wrap around a load as a cover to protect it.
mare mule	A female mule.

maximum registering thermometer	Instrument used to measure body or rectal temperature.
METT-TC	mission, enemy, terrain and weather, troops and support available—time available and civil considerations
mg	milligram(s)
ml	milliliter(s)
mm	millimeter(s)
molly	A female mule.
MOOTW	military operations other than war—Military activities during peace and conflict that do not necessarily involve armed clashes between two organized forces.
MRE	meal, ready to eat
mule	The offspring of a male donkey and a mare (hybrid).
NSN	national stock number
offside	right side of animal
onside	left side of animal
operational continuum	The general states of peace, conflict, and war within which various types of military operations are conducted.
pack string	More than one animal tied together in a file for movement.
pannier	A large container, basket, or bag often carried on the sides and back of an animal.
pcn G	penicillin G
pigtail	Loop of rope used to tie a pack string together during movement.
pkg	package
po	by way of mouth
puller/lead man	The person at the front of a pack string who guides pack animals.
pr	pair
pt	pint
q	every
running W	A type of rope restraint used on both front legs of a horse, allowing a handler to pull out the horse's legs from under it. A practice that should only be done on soft ground to prevent injuring the horse's knees.
SAW	squad automatic weapon
SFOD	Special Forces operational detachment
SO	special operations
SOP	standing operating procedure
TOE	table of organization and equipment

Glossary-3

TOW	tube-launched, optically tracked, wire-guided missile
trail/drag man	Person that observes the pack string from the rear.
travois	A simple vehicle, drawn behind one horse, that consists of two trailing poles serving as shafts and bearing a platform or net for the load or person.
twitch	A loop of rope or chain that is tightened over a horse's lip as a restraining device.
U.S.	United States
USAJFKSWCS	United States Army John F. Kennedy Special Warfare Center and School
UW	unconventional warfare
VEE	Venezuelan Equine Encephalomyelitis
WBGT	Wet Bulb Globe Temperature
WEE	Western Equine Encephalomyelitis
withers	The ridge between the shoulder bones of a horse.
yd	yard(s)
yearling	A year-old equine.

Bibliography

AR 40-905. *Veterinary Health Services*. 16 August 1994.

FM 1. *The Army*. 14 June 2001.

FM 3-05.70. *Survival*. 17 May 2002.

FM 3-07. *Stability Operations and Support Operations*. 20 February 2003.

FM 25-5. *Training for Mobilization and War*. 25 January 1985.

FM 41-10. *Civil Affairs Operations*. 14 February 2000.

FM 55-15. *Transportation Reference Data*. 27 October 1997.

Animal Management. London: His Majesty's Stationery Office, 1923.

Animal Management Transportation. USAJFKSWCS, Fort Bragg, NC, May 1964.

Behind the Burma Road. Dean Brelis and William R. Peers. Atlantic-Little, Brown and Company, 1963.

Care and Feeding of the Horse. Lon D. Lewis. 2d Ed. Williams and Wilkins, 1996.

Cavalry Combat. The Cavalry School, U.S. Army. Norristown: Telegraph Press, 1937.

Drivers Horse Transport. HQ RASC Training Centre. Malaysia, January 1964.

Employment of Pack Animals. Basic School, Marine Corps School. Quantico, VA, November 1961.

Encyclopedia Americana, No. 17. Chicago: Encyclopedia Americana, Inc., 1963.

Encyclopedia Britannica, No. 22 and No. 8. New York: Encyclopedia Britannica, Inc., 1963.

Fundamentals of Riding. Gregor de Romaszkan. New York: Knuph, 1982.

"Guerrilla Warfare in China." James B. Griffith. *The Marine Corps Gazette*. June 1951: 20.

History for 1943 Veterinary Service in North Africa and Mediterranean Theater of Operations. USAMEDS Historical Unit, Office of Surgeon General, Washington, D.C.

Horse Packing in Pictures. Francis Davis. New York: Charles Scribner's Sons, 1975.

Horses, Hitches, and Rocky Trails. Joe Back. Boulder: Johnson Books, 1959.

Manual of Horsemanship, Equitation and Animal Transport. London: His Majesty's Stationery Office, 1937.

Manual of Pack Transportation. H. W. Daly. 1917.

Marsmen in Burma. John Randolph. Houston: Gulf Publishing Company, 1946.

Modern Guerrilla Warfare. Franklin March Osanka, ed. New York: The Macmillan Company, 1962.

"Mules in India." *The New York Times.* New York, 7 June 1964.

"No Plans for Those Mules, Long Ear Type." *The Army Times.* 25 March 1964: 3.

Nutrient Requirements of Horses. 5th Ed. National Academy Press, Washington, D.C., 1989.

Officers Animal Transport Course. HQ RASC Training Centre. Malaysia, April 1963.

Pack Transport and Pack Artillery. Michael F. Parrino. New York: The Queensland Publishing Company, 1956.

Skills of Packing and Shoeing. Vol. 2. Bud Nelson. Jackson Hole, WY, 1975.

The Guerrilla and How to Fight Him. LTC Green. New York: Frederick and Praeger, Inc., 1962.

There Are No Problem Horses, Only Problem Riders. Mary Twelveponies. Boston: Houghton Mifflin, 1982.

Training Video of Special Forces Use of Pack Animals. This video can be purchased at the following web site: http://afishp6.afis.osd.mil/dodimagery/davis/ using the Production Identification Number (PIN): 711656.

United States Army in World War II, China-Burma-India Theater Stilwell's Command Problem. Office of the Chief of Military History, Department of the Army, Washington, DC, 1956.

U.S. Fifth Army History. 7 October to 15 November, Part II, A Volturno to the Winter Line.

"Where the Troops Are." LTC Frank F. Rathburn. *Army.* March 1964: 39.

X-File 3-35.23. *Small Wars Animal Packers Manual.* Marine Corps Warfighting Laboratory (MCWL) Training Command and Mountain WarfareTraining Center (MWTC), 3255 Meyers Avenue, Quantico, VA. 17 November 2000.

Index

A
animal conformation, 2-5 through 2-9

B
barrel hitch, 7-20
basket hitch, 7-21
bivouac, 8-15, 8-16
bridle, 6-1, 6-2
burns, 4-12

C
camels, 10-5, 10-6
cinching, 5-14, 5-15
combat considerations
    equipment, 6-16
    weapons, 6-14, 6-15
cover and concealment, 9-2

D
detachment personnel, 8-1, 8-2
diamond hitch, 7-22
dogs, 10-6, 10-7
donkey characteristics, 2-2 through 2-4

E
elephants, 10-8
euthanasia, 4-24, 4-25

F
farrier tools, 3-5, 3-6
feeding, 2-14 through 2-19
    field, 2-24
    garrison, 2-23

first aid
    supplies, 4-7 through 4-9
    treatment, 4-9 through 4-11

G
grazing, 2-22
grooming, 3-1 through 3-3

H
halter, 5-5, 5-6
harness, parts of, 5-11
high line, 7-26
hobbles, 2-10

I
immunization, 4-22

K
knots, 7-2, 7-3

L
lameness, 2-13
litters
    suspended, 7-31
    travois, 7-30
llamas, 10-1 through 10-5

M
medical supply list, 4-23
mission, enemy, terrain and weather, troops and support available–time available and civil considerations (METT-TC), 1-2, 1-3
movement procedures, 8-2, 8-3
mule characteristics, 2-1

N
night horse, 7-28, 7-29

P
packsaddles, 5-1 through 5-4
pack string, 7-23 through 7-25
pharmacological listing, 4-24
picketing, 7-27

R
rations, types of
    emergency, 2-18
    field, 2-18
    garrison, 2-18
restraints, 3-18 through 3-21
riding techniques, 6-9 through 6-11

S
saddles
    McClellan, 6-5
    Western, 6-4
saddle, fitting of, 5-6 through 5-8
security, 9-1
shoeing
    front feet, 3-6, 3-7
    hind feet, 3-8, 3-9
slings and hitches, 7-18 through 7-22

U
urban environments, 9-3

W
water crossing, 8-5 through 8-9

FM 3-05.213 (FM 31-27)
16 JUNE 2004

By Order of the Secretary of the Army:

PETER J. SCHOOMAKER
General, United States Army
Chief of Staff

Official:

JOEL B. HUDSON
Administrative Assistant to the
Secretary of the Army
0415302

DISTRIBUTION:

Active Army, Army National Guard, and U. S. Army Reserve: To be distributed in accordance with initial distribution number 115753, requirements for FM 3-05.213.

www.ingramcontent.com/pod-product-compliance
Lightning Source LLC
Chambersburg PA
CBHW071206240526
45470CB00018B/1515